# Income stabilisation in European agriculture

# Income stabilisation in European agriculture

## Design and economic impact of risk management tools


edited by:
**Miranda P.M. Meuwissen**
**Marcel A.P.M. van Asseldonk**
**Ruud B.M. Huirne**

*Institute for Risk Management in Agriculture*
*Wageningen University*
*The Netherlands*



Wageningen Academic Publishers

SIXTH FRAMEWORK PROGRAMME

**Disclaimer**

This publication does not necessarily reflect the European Commission`s views and in no way anticipates the Commission`s future policy in this area. Its content is the sole responsibility of the authors. The information contained herein, including any expression of opinion and any projection or forecast, has been obtained from or is based upon sources believed by the authors to be reliable but is not guaranteed as to accuracy or completeness.



**ISBN 978-90-8686-079-1**

**First published, 2008**

**Wageningen Academic Publishers**
**The Netherlands, 2008**

# Table of contents

# Contributors

**Berg, Ernst**, Department of Farm Management, Rheinische Friedrich-Wilhelms-Universität Bonn, Meckeneheimer Allee 174, 53115 Bonn, Germany, e.berg@uni-bonn.de

**Bielza, Maria**, Agricultural Unit, Joint Research Centre, Via E. Fermi 2749, 21027 Ispra (VA), Italy, maria.bielza@jrc.it

**Dalton, Graham E.**, Dalton Associates, Nethercontlaw, Miltimber, Aberdeen, UK, graham@contlaw.com

**Garrido, Alberto**, Department of Agricultural Economics and Social Sciences, Avda. Complutense s/n, 28040 Madrid, Spain, alberto.garrido@upm.es

**Guba, Waldemar,** Department of Business Economics and Organization, Warsaw University of Life Sciences, Nowoursynowska 166, 02-787 Warsaw, Poland, wguba@poczta.pl

**Huirne, Ruud B.M.**, Institute for Risk Management in Agriculture, Wageningen University, Hollandseweg 1, 6706 KN Wageningen, The Netherlands, ruud.huirne@wur.nl

**Kramer, Joern**, Department of Farm Management, Rheinische Friedrich-Wilhelms-Universität Bonn, Meckeneheimer Allee 174, 53115 Bonn, Germany, joern.kraemer@uni-bonn.de

**Landmesser, Joanna**, Department of Business Economics and Organization, Warsaw University of Life Sciences, Nowoursynowska 166, 02-787 Warsaw, Poland, joanna_landmesser@sggw.pl

**Majewski, Edward**, Department of Business Economics and Organization, Warsaw University of Life Sciences, Nowoursynowska 166, 02-787 Warsaw, Poland, edward_majewski@sggw.pl

**Meuwissen, Miranda P.M.**, Institute for Risk Management in Agriculture, Wageningen University, Hollandseweg 1, 6706 KN Wageningen, The Netherlands, miranda.meuwissen@wur.nl

**Palinkas, Peter**, Department of Farm Economics and Management, Szent István University, Gödöllő, H-2103 Gödöllő, Páter K. u. 1., Hungary, palinkas.peter@gtk.szie.hu

**Poppe, Krijn J.**, Agricultural Economics Research Institute, Alexanderveld 5, 2585 DB The Hague, The Netherlands, krijn.poppe@wur.nl

**Székely, Csaba**, Department of Farm Economics and Management, Szent István University, Gödöllő, H-2103 Gödöllő, Páter K. u. 1., Hungary, szekely@gtk.gau.hu

**Van Asseldonk, Marcel A.P.M.**, Institute for Risk Management in Agriculture, Wageningen University, Hollandseweg 1, 6706 KN Wageningen, The Netherlands, marcel.vanasseldonk@wur.nl

**Vrolijk, Hans C.J.**, Agricultural Economics Research Institute, Alexanderveld 5, 2585 DB The Hague, The Netherlands, hans.vrolijk@wur.nl
**Was, Adam,** Farm Management, Szkola Glowna Gospodarstwa Wiejskiego, Nowoursynowska 166, 02-787 Warsaw, Poland, adam_was@sggw.pl

## Steering committee of Income Stabilisation project

**Burgaz, Fernando**, ENESA, Spain, fburgazm@mapya.es
**Cafiero, Carlo**, Napoli University, Italy, cafiero@unina.it
**Fonk, Jan**, Interpolis Achmea, The Netherlands, jan.fonk@interpolis.nl
**Gargano, Nadia**, European Commission, DG Agri, Belgium, nadia.gARGANO@ec.europa.eu
**Georgiadis, Nikiforos**, ATEbank, Greece, ngeorgiadis@ate.gr (nikge@yahoo.com)
**Hartell, Jason**, GlobalAgRisk, Kentucky, USA, jason@globalagrisk.com
**Kasten, Erich**, Agriculture Reinsurance Consultants GmbH Switzerland & AgroRisk Ukraine Ltd. Ukraine, dr.e.kasten@bluewin.ch
**Koeckhoven, Bernard**, Interpolis Achmea, The Netherlands, bmw.koeckhoven@interpolis.nl
**Lutzeyer, Hans-Joerg**, European Commission, DG Research, Belgium, Hans-Joerg.Lutzeyer@ec.europa.eu
**Rasmussen, Svend**, The Royal Veterinary and Agricultural University Copenhagen, Denmark, svend.rasmussen@flec.kvl.dk
**Schulz-Greve, Willi**, European Commission, DG Agri, Belgium, willi.schulz-greve@ec.europa.eu
**Skees, Jerry**, University of Kentucky & GlobalAgRisk, Kentucky, USA, jerry@globalagrisk.com
**Suarez, Carmen**, NFU, UK, carmen.suarez@nfu.org.uk
**Wilkens, Eckhardt**, Committee European d'Assurance (CEA), Germany, wilkens-burgwedel@t-online.de

# Acknowledgements

The editors and contributors gratefully acknowledge financial support from the European Commission, under the European Union's Sixth Framework Programme for Research and Technological Development, for the project 'Income Stabilisation.'

Authors also thank constructive comments and suggestions from the steering committee members throughout the full course of the project. Their practical insights reflecting various stakeholders and nationalities were useful – and livened up our project consortium meetings. Also project meetings in Brussels were highly appreciated.

In addition, participants of the 2008 workshop in Warsaw are acknowledged for contributing with respect to finetuning our project results to Eastern EU agricultural circumstances. Similarly, scientific reflections from contributors to the 108th seminar of the European Association of Agricultural Economists, entitled 'Income stabilisation in a changing agricultural world: policy and tools,' were greatly appreciated.

# Abbreviations

| | |
|---|---|
| CAIS | Canadian Agricultural Income Stabilisation program |
| CAP | Common Agricultural Policy |
| EU | European Union |
| FADN | Farm Accountancy Data Network |
| NISA | Net Income Stabilisation Account |
| WTO | World Trade Organisation |

## Member states of the EU

| | |
|---|---|
| AT | Austria |
| BE | Belgium |
| BG | Bulgaria |
| CY | Cyprus |
| CZ | Czech Republic |
| DE | Germany |
| DK | Denmark |
| EE | Estonia |
| EL | Greece |
| ES | Spain |
| FI | Finland |
| FR | France |
| HU | Hungary |
| IE | Ireland |
| IT | Italy |
| LT | Lithuania |
| LU | Luxembourg |
| LV | Latvia |
| MT | Malta |
| NL | Netherlands |
| PL | Poland |
| PT | Portugal |
| RO | Romania |
| SE | Sweden |
| SK | Slovakia |
| SI | Slovenia |
| UK | United Kingdom |

# Chapter 1

# Income stabilisation in agriculture; reflections on an EU-project

*Miranda P.M. Meuwissen, Marcel A.P.M. van Asseldonk and Ruud B.M. Huirne*

## 1.1 Background

### *1.1.1 Changing risks*

Income from farming is rather volatile due to stochastic (random) factors that affect production and prices. Throughout the years, various risk management tools have been used to reduce, or to assist farmers to absorb, some of these risks. Also the Common Agricultural Policy (CAP) of the European Union has taken away some of the risks through a variety of mechanisms that support prices of many agricultural products. Price and production risks, however, are likely to change in the near future. International trade agreements can be expected to lead to price liberalisation and to more exposure of European farmers to competitive market forces. Production risks may change, among others, due to a more regulated use of herbicides, medicines and vaccines. Changes are also occurring in relation to risks of catastrophic events such as floods and livestock epidemics. In the past, farmers have often been compensated for such losses by governments but there is increasing pressure to find more private market solutions, or at least more formal public-private arrangements.

### *1.1.2 Limited agricultural risk markets in Europe, in contrast to the USA and Canada*

In the European Union, at present there are only limited agricultural risk markets to deal with price and production risks. For example, in most member states, arable farmers can insure against loss of crop production caused by only a few specified perils. For livestock farms there is usually no low-yield insurance, but it is generally possible to insure against stock mortality from specific accidental causes. Price risks can be managed to some extent by forward contracting or by hedging on futures markets. The latter option, however, is available for very few commodities and can also be subject to considerable basis risk.

In contrast to Europe, in the USA and Canada a myriad of new agricultural insurance products and income stabilising tools has emerged. These include enhancements on traditional crop insurance, revenue insurance, farm accounts, whole-farm insurance, and even subsidies on price options for milk (Kennedy and Yang, 1982; Goodwin and Ker, 1998; Gundersen *et al.*, 2000; Barnett *et al.*, 2005; RMA, 2008). In Canada, farmers could, until 2002, participate in the Net Income Stabilisation Account (NISA), see for instance Stokes *et al.* (2000). In 2003, this program was replaced by the Canadian Agricultural Income Stabilisation (CAIS) program. This program integrates stabilisation and disaster protection into a single program, helping producers protect their farming operations from both small and large drops in income (Dismukes and Durst, 2006).

### *1.1.3 Considerations within the EU*

Since 1998, the European Community has been intensively investigating the potential role of agricultural insurance programs in stabilising agricultural incomes (Meuwissen *et al.*, 1999; OECD, 2000a; European Commission, 2001; Conference on risk management and agricultural insurance in the European Union by the Spanish Presidency in May 2002; Conference on natural risk and insurance in the agricultural sector, perspectives in the Europe of 25 by the Greek Presidency in June 2003; Meuwissen *et al.*, 2003; Cafiero *et al.*, 2005; European Commission, 2005; European Commission, 2006a). World Trade Organisation (WTO) agreements and the EU enlargement mean that agricultural policy in Europe is also evolving. As the insurance schemes and income stabilising tools developed in the USA and Canada seem legitimate in the WTO-framework, i.e. they fit into the 'green box' representing allowed forms of support, the European Community might also consider alternative risk financing tools. Moreover, in 2008, following the latest 'CAP health check', the European Parliament recommended, among others, to replace intervention systems for market crops by a safety net for crises, as well as to develop private sector or mixed insurance schemes, and to consider a Community-wide reinsurance system for climate-related or environmental disasters (European Parliament, 2008).

## 1.2 Scope and objectives

The Income Stabilisation project analysed the opportunities that risk management tools offer for stabilising farm incomes in the European Union in the contemporary context of new agricultural risks, an enlarged European Union, changing views about eligible forms of income support and disaster relief, and on-going developments at international risk management markets.

The project had three pillars: (1) a detailed analysis of farmers' *risk exposure* in the past and projected risks in the future; (2) a review of *risk management experience and farmers' perceptions of risk*; and (3) the *economic impact of and policy options for viable risk management instruments*. The scope of the project is reflected in Table 1.1.

**Table 1.1.** *Scope of Income Stabilisation project.*

| | |
|---|---|
| Risk | Normal business risk and crisis risk[1] |
| Farm types | Specialised and mixed farms, large-scale and small-scale farms |
| Commodities | Crop and livestock |
| Income | Farm level, not household level |
| EU regions | Established and new member states[2] |
| Risk management | Private, public and public-private partnerships |

[1]Including single farm and multiple farm crisis risk, and 'foreseeable' and unpredictable crisis risk.
[2]In-depth analyses focus on participating member states: Germany, Hungary, the Netherlands, Poland, Spain.

### *1.2.1 Risk exposure*

#### Past risk exposure: let farm data tell the story

This component of the project thoroughly analysed farm-level data from the Farm Accountancy Data Network (FADN) in order to provide insight into the volatility of farmers' risk exposure and into the chances of facing catastrophically low production, price or income levels. Specific research objectives included: (1) to provide insight into farm-level production, price and income distributions, including downside risk; (2) to make a clear distinction between normal income fluctuations and income crises; and (3) to discuss the usefulness of FADN data for measuring farmers' income (crisis) risk.

### Future risk exposure: what can we expect in the next ten years?

Since risks are constantly changing, data from the past generally tell only *part* of the story. This component of the project provided insight into the impact of future World Trade Organisation (WTO) and CAP scenarios on farmers' risk exposure and risk management opportunities. Research objectives were (1) to define likely future CAP and WTO scenarios with their implications for price, production and farm income; (2) to analyse the impact of these scenarios on price, production and income probability distributions of farmers in the European Union; and (3) to analyse the impact of these scenarios on the chances of catastrophically low incomes.

## *1.2.2 Risk management experience and perception*

### Review of the international risk management arena

This part of the project reviewed historical, current and developing risk management instruments, both within the European Union and in non-EU countries. Research objectives were (1) to report on successful and unsuccessful risk management instruments; and (2) to analyse the major characteristics of these schemes (e.g. the risks covered and the underwriting criteria applied), and their performance and economic impact. Important performance issues are loss ratios and participation rates, while the relevant economic criteria are the degree of risk reduction achieved, the reported incentive issues for all parties involved, and the budgetary implications.

### Risk and risk management perception: an investigation of farmers' beliefs

Risk analyses by scientists do not necessarily correspond to the perception of risks by farmers. Also, what may be perceived as theoretically promising risk management instruments may not work well for farmers. What is needed, then, is an in-depth analysis of farmers' perceptions of risk exposure, and (new) risk management instruments. Research objectives of this component of the Income Stabilisation project were: (1) to analyse farmers' perceptions of (crisis) risk and (crisis) risk management; and (2) to analyse farmers' perceptions of the role of various possible risk financing partners, ranging from their own role to that of national and European governments.

### *1.2.3 Economic impact and policy options*

The economics of risk management instruments

Theoretically ideal concepts may not work well at the farm level, or may just be too costly. This component of the project modelled the economic impact of potential risk management instruments for the European Union. Research objectives included: (1) to select promising risk management instruments, including purely private instruments, public-private partnership instruments and entirely public risk management instruments; (2) to develop a whole-farm optimisation model and to analyse the economic impact of potential risk management instruments at farm level, including their impact on production decisions and on the level of (crisis) risk; and (3) to analyse the budgetary impact of potential risk management instruments.

Policy options for risk management

The various themes of the project, i.e. risk exposure, risk management experience, risk perception and the economic impact of various schemes, were brought together in the final component of the project. The main research objective was to synthesise all previous project components in order to come up with policy options for (crisis) risk management that are feasible from a design and budgetary point of view, legitimate in CAP and WTO frameworks, and interesting for farmers.

## 1.3 Major findings and reflections on the EU-project

### *1.3.1 Risk exposure*

Past risk exposure

> *Main conclusion: Data from FADN can partly provide insight into the wide range of on-farm risks. (Chapter 2)*

1. Farmers are confronted with a wide range of factors that affect their income. Besides a continuing increase in productivity, fluctuations in yields due to climatic conditions and fluctuations of prices of outputs and inputs strongly affect the levels of incomes. Contagious diseases affecting the production of crops and animals are external events that can cause a crisis for a single farmer or a group of farms.
2. The analyses of individual farm data showed strong fluctuations in farm income. Large differences were found between different member states, regions

and sectors. Furthermore, analyses showed strong differences between farms within the same type of farming. The size of a farm only explains a small part of these differences. Furthermore, average farm incomes showed to only convey a limited amount of information. On the one hand, average incomes do not show that even in case of a positive average there can be a large group of farms with low or even negative incomes. On the other hand, the strong fluctuations of incomes and the strong changes in the relative income position of farms stress the importance to look at multiple year averages to draw meaningful conclusions on the level of income and the standard of living of individual farmers.

3. The availability of information on off-farm income is still limited, especially in the FADN framework, but there are clear indications that the importance of off-farm income is increasing. Off-farm income is more stable than farm income and thus provides a cushion for farm income fluctuations. Also, off-farm assets are essential in understanding farmers' behaviour and their ability to cope with crises. Therefore, current data sets do not allow to predict whether a crisis will lead to bankruptcy or whether sufficient resources are available to absorb a shock at household level.
4. The analyses also showed that there are large differences in the shortfall risk of farms, i.e. farms' risk of incomes below zero. Simulated crises show that farms in North-Western Europe have a much higher shortfall risk due to the structure of farming. Small margins lead to much more difficulties in absorbing a shock in the short run. Case descriptions illustrated that market response to certain events can increase income fluctuations. For instance, market response after classical swine fever in the Netherlands led to a strong increase in production due to temporary higher prices and therefore to a collapse of prices and farm incomes.
5. Although FADN data can clearly show the impact of factors such as heavy rainfall and classical swine fever on farm incomes, a clear distinction between normal income volatility and income crises due to external events is difficult to derive. FADN data reflect an aggregation of all events happening on a farm during a year. As year to year fluctuations are relatively strong, further information is needed to establish whether a strong effect is caused by an external crisis such as heavy rainfall, or by other circumstances such as bad management or illness of the farmer.

### Future risk exposure

*Main conclusion: Increasing levels of liberalisation do not seem to lead to widespread negative effects on the stability of farm-incomes in the*

*EU, although vulnerability differs significantly across farm types and member states. (Chapter 3)*

1. Results showed the considerable dependence of incomes and income variability on CAP support. Dependence however varied greatly across farm types, size and member states. Variation is caused by differences in terms of production patterns, changes in support levels, farming sector structures, inputs used and differences in the volatility of commodity prices.
2. Simulation results showed that at least part of the agricultural sector is relatively immune to further liberalisation of market and farm support policies (scenarios 2018). Given relatively positive prospects for the future world market situation there is no significant change in income levels compared to the basis and the most likely 2013 scenario, provided that direct support is maintained. Furthermore, in a likely 2018 scenario which is more protective, farm incomes on most farm types in all member states improve, largely due to predicted productivity increases. Reductions in direct support (greater modulation and ceiling) assumed in another 2018 scenario give more visible decreases in incomes, but still not very different from the base situation.
3. A complete market liberalisation did not lead to widespread negative results for EU farming, although selected farm types in some of the member states analysed would be affected stronger than others. An adverse effect of liberalisation is the greater risk of low incomes as measured by the percentage of farms with a negative farm income. This is caused by both the lowering of agricultural commodity prices due to diminishing price support and increased variability of yields and prices as assumed in the model. This finding applies mainly to the scenarios in which agricultural subsidies were significantly reduced. Under the protectionist scenario, which assumed a return to a high support, Agenda 2000 type of agricultural policy, the risk of low incomes is greater than in the base scenario. This is explained by increased variability but also by unfavourable trends in price and cost developments over a long-term period.
4. The increase in the risk of low incomes under more liberal scenarios varies between farm types, depending on production type and economic size. The level of risk on specialised dairy and pig farms is less significant than on crop farms. This is likely due to the relative stability in productivity of livestock and the limited exposure of animal farms to market related risks.
5. There are some indications that the risk of low incomes significantly increases on some of the smallest farm types (8-16 ESU) and also on large farms, i.e. above 100 ESU.
6. Any common income stabilisation scheme for the EU as a whole will need to consider that farms' vulnerability differs across member states. Income variability and the risks of loss were shown to increase, in some cases even quite

dramatically, with increasing levels of liberalisation but to noticeably different degrees. The lesson seems to be that there are internal strategies for farms of all types to moderate any increased price risks including structural adjustments. The danger is that the introduction of income stabilisation measures might slow down the adoption of such practices. Perhaps somewhat surprisingly there are some farm types, especially in Poland and Spain, where even the most liberal scenarios do not produce much reduction in income levels or much greater risks of negative incomes.

### *1.3.2 Risk management experience and perception*

#### Review of the international risk management arena

> *Main conclusion: There is an increasing variety of risk management instruments, including tools to capture the problem of asymmetric information. Public sector involvement can (still) lead to undesired incentives. (Chapter 4)*

1. Contingent-state contracts, futures, options and other index derivatives are useful mechanisms to manage on-farm risks. However, agriculture in the EU is extremely diverse under natural conditions, as in terms of risks and structural situations. Widely traded securities that permit hedging risks will be difficult to develop, because basis risks and trading costs will be a serious obstacle for take-up by farmers. Yet, as technological innovations enable the development of more diverse index instruments, a market may develop for these in the EU. At present, the use of financial instruments among farmers and even cooperatives is low.
2. Farmers would profit from a diversified set of risk-management instruments that target multiple risk sources both within-farm boundaries and across the market chain from the farm-gate to the wholesale market. In highly capitalised agriculture we are seeing major innovations in contractual agreements along the market chain that will enable professional farmers to externalise part of their risks.
3. When risk instruments are subsidised, it is a general rule that instruments with higher coverage and risk reduction potentially come with lower subsidy efficiencies. More euros are needed in relative terms to increase risk reduction effects, once these are already large. Yet, in the case of insurance, reducing subsidies would likely be followed by lower rates of use of instruments.
4. OECD countries seem to have developed two alternative models to provide safety nets and risk management tools to their farmers. The keywords of the so-called 'Model 1' are: training, competitiveness, liberalisation and compensation

schemes for catastrophes and crises. For 'Model 2' these are crop and livestock insurance, premium subsidies, gradual reduction of public compensation and increasing importance of insurance. In the EU, 'Model 1' seems to be followed by Northern member states, whereas 'Model 2' is generally the approach of Mediterranean countries, although Austria's policies fit better with 'Model 2'. These two models cannot easily converge to a middle ground mix. At most, member states are increasingly requiring that farmers contract insurance to become eligible for *ad hoc* compensation payments in case of a crisis.

5. Actuarial loss ratios of mature and growing agricultural insurance systems in the world have shown consistency and soundness. Actuarial techniques improved significantly, enabling countries to control problems of asymmetric information and poor loss adjustment procedures. Technologies, data mining, surveillance, and better risk evaluations explain these improvements. The era of poor insurance performance indicators around OECD countries came to an end in the mid-1990s.
6. Publicly provided (or publicly regulated and subsidised) crop insurance in OECD countries suffered from problems of asymmetric information in the 80s and early 90s. Current loss ratios of private insurance do not seem to differ significantly from those of publicly provided insurance. Also, on the EU scale, actuarial ratios do not seem to differ significantly among member states with or without subsidised premiums. Note hereby that 'loss ratios' were calculated as indemnities paid divided by premiums received, whereby premiums received include both farmers' premiums and (if applicable) government subsidies. The non-significant differences found therefore imply negligible differences in transaction costs between privately and publicly provided schemes.
7. OECD countries are constantly innovating and developing new instruments to underwrite or transfer risks and to provide new guarantees. Many of these are technology-based and have great potential because administrative costs are much lower than in traditional crop insurance. Innovations are also of institutional nature, like new contractual definitions and design, market regulations and new modes of government participation.
8. Growing insurance portfolios increase the effects of risk-pooling and reduce the cost of reinsurance in relative terms. Some hazards, such as droughts or epidemics, for which disaster payments are offered in some countries, are now insurable, even though in most cases with backup of some sort of public reinsurance. In the near future, the trade of weather derivatives in the derivative markets can further increase this effect and hopefully permit private insurance of systemic risks.
9. While many working documents differentiate between normal risks and crises/ catastrophes, past and existing policies cannot be equally categorised. There

are countries whose definition of catastrophes encompasses hazards that are considered normal risks by others, and vice versa.

### Risk and risk management perception

> *Main conclusion: Risk perception varies considerably across member states. Price and weather risk however is generally perceived to be most threatening. (Chapter 5)*

1. When considering factors that affect farming, it became clear that farmers in the studied member states and also in the different economic size and farming activity groups perceive weather and natural disasters and price volatility the elements having the largest effects on their farming. It strongly corresponds to the finding that climatic and market conditions were the primary reasons of critical situations experienced by farmers.
2. The range of instruments applied by farmers to manage risks related to agriculture show that specific crop and livestock insurance is widespread mainly in the established member states while property insurance has an important role in both the established and new member states. Although, the established member states have well developed financial markets, a high proportion of farmers in all selected member states tries to avoid using credit and thus taking on new liabilities. A widespread way to secure sufficient financial resources is to hold financial reserves, especially in the new member states and Germany. This method is reasonable in new member states where the majority of farmers perceives the adequacy of access to credit in a less positive way than those in the established member states.
3. When future use of risk management instruments is considered, it becomes quite clear that the majority of farmers perceives current instruments as adequate for offsetting risks, so they are going to continue to apply them. The same applies when considering larger farms, where a majority of farmers would apply the currently used methods also in the future, although with the increase of farms' economic size, farmers are more interested in hedging (futures and options), which is a sophisticated method of risk management.
4. A special way of dealing with risks related to agriculture is to find work outside farming and generate revenues additional to farming as to offset the volatility of farm incomes by other income sources. In the established member states a higher proportion of farmers have off-farm revenues (except the Netherlands where farming provides sufficient earnings) than in the new member states where there are no job opportunities outside farming in rural areas. However, where off-farm revenue is available it is generated continuously throughout the year in the majority of cases, both in established and new member states.

5. An important aspect of risk for farmers is to see how farmers perceive the EU and national institutions and regulatory changes related to farming. From the analyses it is concluded that farmers do not attribute high risks to these institutions and the adaptation process to changes in regulations, however there still seems to be space to improve farmers' acceptance and opinions both on the EU and national levels.
6. A major finding of the perception study is that, although high similarities were expected between pairs of member states such as Germany-Netherlands and Hungary-Poland in terms of farmers' risk perception and risk management strategies, it is concluded that such similarities only exist at the level of individual, highly specific issues; at the general level many differences were found among these member states. The case of Spain is an outstanding example on how complex the network of similarities and differences can be – as Spain is very close to other member states in some respects, while completely different for other issues. Due to the large amount of differences found, it is argued that the adequate answer for the risk management challenges at the EU level could lie in the establishment of a flexible risk management policy framework which could be well customised based on the specific needs of individual member states while meeting universal guidelines across the Union.

### *1.3.3 Economic impact and policy options*

#### The economics of risk management instruments

> *Main conclusion: Despite application of a portfolio of prospective risk management instruments, substantial on-farm income volatility remains. Also, on-farm diversification seems to have its limitations. (Chapter 6)*

1. Farm risk exposure differed between the assumed future policy scenarios substantially. The pattern of changes in the level of expected farm income across scenarios is similar for the five case farms under investigation. On the long run expected farm incomes increase under more protectionist policies but are depressed if liberalisation is assumed.
2. The impacts of alternatively policy scenarios on the optimal farm plan were not substantial. The optimal farm plan of general field cropping farming systems as well as specialised cereals, oilseed and protein farms is marginally altered. The amount of cash crops cultivated – which are characterised by higher but more volatile outcomes – is more affected by agronomic constraints rather than future policy scenarios.

3. Diversification as an on-farm risk management tool has its limitations. The analysis of the case-specific trade-off between the expected gross margins and risk provided an indication of the efficiency of farm diversification. This is to say when decisions are made assuming risk neutrality or moderate risk aversion whereby farmers are not willing to forego a substantial part of the expected income in order to avoid the risks associated with the cultivation of more risky cash crops.
4. Substantial volatility remains despite prospective risk management instruments considered. Farming is in general a risky business since crop yields and prices are relatively volatile in comparison to the expected farm income. In conjunction with a strike level set at for example 80% of the expected outcome, implying a deductible of 20%, explain the riskiness.
5. The budgetary implications of prospective risk management tools differed between case farms and future policy scenarios under consideration. To address the implications for budgetary expenses the payments involved with insurance were studied. However, the budgetary consequences stemming from insurance depend on the level of granted premium subsidy. Total payments per hectare were affected by the alternatively policy scenarios. Gross premium payments on general field cropping farming systems, with more volatile cash crops, exceeded those on specialised cereals, oilseed and protein farms. Expected gross premiums under more liberal policies were more substantial than under protectionist policies offsetting the decoupled payments.

## Policy options for risk management

> *Main conclusion: For normal on-farm enterprise risk, infrastructural improvements are needed. For crisis risk, rules need to be set at EU level, but premium subsidies should be avoided. (Chapter 7)*

With regard to *normal on-farm enterprise risk* there is a need for public policy measures intending to facilitate the operation of private markets, including insurance as well as other financial instruments. More precisely such measures could aim at:

1. Education of farmers and extension personnel in risk management issues, particularly in the functioning and the use of derivative markets.
2. Support of the development of private insurance/derivative markets, e.g. index-based insurance or weather derivatives, without paying premium subsidies. Support may include (1) providing the regulatory institutions and informational support; (2) the development of informational infrastructure such as monitoring equipment and databases; (3) direct participation in the market during the starting phase, e.g. by offering options based on weather

indexes, or by providing public re-insurance; and (4) other forms of start-up support.
3. Support of the development and operation of mutual funds. Public policy could provide matching contributions to those of the farmers and set up the rules for funds' withdrawals. This can be a viable option to securitise production risks in the case of specialty crops or animal diseases.
4. Support of institutional arrangements in the sense of public private partnerships that provide risk management services to the farmers. Since the selection of an optimal portfolio of risk management tools is a complex task it can be doubted that farmers – besides all other tasks they have to accomplish in their predominantly small to medium sized operations – will ever be able to cope with this problem. Instead, special institutions could take over the task of creating and managing such portfolios that fit the need of particular farm types. The farmers themselves would then only have to deal with one aggregate instrument aimed at reducing their downside risk of income.

With respect to on-farm consequences of *crisis risk* there are basically two options for public policy:
1. Direct damage compensation after the event has occurred is the only option in the short term. Following Cafiero *et al.* (2005), only damages to farm assets, such as buildings, equipment, green-houses, perennial crop stands and breeding livestock, should be directly compensated. Also, rules are set at EU level stating the conditions under which disaster relief will be granted, i.e. type of event, extent of losses, and proportion of the loss that is compensated.
2. In the medium and long term there is also the possibility for preventive actions like public investments in protective infrastructure or the support of private actions that reduce the extent of damages caused by disastrous events. Presently, the world food system operates at historically low inventory levels and is therefore highly vulnerable to supply shortages due to bad weather conditions or other catastrophic events. In this situation, public investments in buffer stocks could be a valuable risk management policy. Preventive actions might also include measures that aim at establishing viable private markets for catastrophe insurance. However, premium subsidies would not be included in this set.

## 1.4 Reflections from an Eastern EU perspective

### Finetuning project results to Eastern EU agricultural circumstances

> *Main conclusion: Project findings are generally considered to be applicable to Eastern EU agricultural circumstances. (Appendix A)*

1. Experts generally agree on project conclusions to be applicable to Eastern EU agriculture. Strongest agreement, both for experts originating from Eastern EU member states and non-Eastern EU members states, was found for the need for public policy to facilitate private markets by educating farmers and extension workers in risk management issues and the use of derivate markets. Experts also agree on crises to be mainly caused by weather and market risks. They furthermore agree that some member states are more exposed to income falls than others and (as a consequence) that risk premiums per hectare differ substantially across member states. There is also agreement on decoupled direct payments playing a key role in stabilising farm incomes under a less protective CAP. Experts in addition agree that many innovations, such as derivatives and public-private risk sharing, enlarge the opportunities for transferring risks, although experts originating from non-Eastern EU member states seem to be more confident about this issue.
2. For some project findings, experts doubted their general applicability to Eastern EU agricultural circumstances. This is for instance the case with regard to the expected farm-level impact of future scenarios. Experts, especially those not originating from Eastern EU member states, do expect a significant impact from WTO agreements on farm incomes. In addition, they believe that more liberal policies actually will induce arable farmers in Eastern EU member states to change their farm plans. They furthermore doubt whether mostly crop farms and small and large farms face income risk. Moreover, 50% regards diversification as potentially becoming an important risk management tool in Eastern European agriculture.
3. With regard to some more general statements on income stabilisation issues, experts to a large extent agree that farmers are not fully aware of the wide spectrum of risk management tools available. They also agree that insurance schemes for production risks are underdeveloped and that insurance schemes are not only attractive in case of subsidies or catastrophic events. The latter issue, i.e. the wider applicability of insurance schemes, was especially stressed by experts from non-Eastern EU member states.

## 1.5 Conclusions and outlook

The Income Stabilisation project aimed at analysing the opportunities of risk management tools for stabilising farm incomes in the European Union. Results present a representative cross-section of European farming, i.e. risk exposure, risk perception and economic impact issues have been studied in depth for 5 member states, i.e. Germany, Hungary, the Netherlands, Poland and Spain. General conclusions from the project include:

1. There is not sufficient data to accurately and instantly assess farmers' income risk. FADN data are worthwhile for farm-level risks, but only partly and retrospectively. Off-farm income data are largely lacking.
2. Income risk measures and methods used throughout the analyses, i.e. income variation at individual farm level, shortfall risk and whole-farm modelling, provide insight into the wide variety of income risks faced by EU farmers.
3. Eastern EU finetuning among stakeholders disclosed that there is not a commonly shared risk management perception, knowledge and view on the prospects of (innovative) risk management solutions.
4. Due to established differences among EU-farming, both in risk exposure, risk perception and whole-farm consequences, risk management solutions need to be 'tailor-made'. However, from an analytical perspective, premium subsidies are nowhere to be preferred. Moreover, all risk management instruments require careful and proper design.

From the experiences from the project, also a number of issues for further research have been identified:
1. Public sector involvement in on-farm risk management. If, from a political perspective, this option is considered, further research is needed to transparently set the rules for support and design. Rules should aim at public involvement leading to 'crowding *in* private markets' instead of 'crowding *out*' these markets.
2. Crisis risk assessment. As FADN data did not fully allow to analyse farm-level effects of crisis risk, further analyses should be directed towards specific risks such as liability risk and disruptions through environmental damages.

# Chapter 2

# Income volatility and income crises in the European Union

*Hans C.J. Vrolijk and Krijn J. Poppe*

## 2.1 Objectives

Farmers' incomes is a theme of interest for policy makers, the press, the general public and above all farmers. Much attention is paid to the structural development and trends in the income levels; less attention has been paid to fluctuations in incomes. Farmers' incomes show strong fluctuations over time due to inherent fluctuations in prices and yields. Fluctuations in yields mainly originate by natural conditions such as draught, heavy rain, frost and animal diseases and often lead to even stronger price fluctuations (in non-regulated markets). Furthermore, farm incomes differ strongly within countries even among farms of the same farming type and farm size. External events directly affect farm incomes but also have strong indirect effects due to market reactions. This chapter focuses on the analyses of FADN data to develop an understanding of the volatility of farm incomes, prices and yields in agriculture. The objectives of this chapter are:

- To analyse individual farm data with respect to price, production and farm income (FADN).
- To complete farm analyses with data on farm households, where possible (OECD, ERS).
- To provide full insight into the price, production and income distributions including downside risk.
- To make a clear distinction between normal income fluctuations and income crises.

Volatilities of prices and yields have been calculated for other chapters in this project. The analyses described in this chapter mainly focus on the volatility of farm incomes.

Section 2 will describe previous results on farm income volatility and the impact of off-farm income on the volatility of incomes. Section 3 describes the data source and the methodology applied. Section 4 gives a short description of some major structural trends in agriculture to provide the background for understanding the findings with respect to volatility. Section 5 gives a detailed analysis on the volatility of farm income. Section 6 investigates income crises and shortfall risks. The chapter ends with conclusions on the volatility of farm incomes and on the usefulness of FADN data as the major source to analyse farm incomes.

## 2.2 Previous research on farm income and volatility

Farm income is affected by numerous factors. Technical development results in increasing levels of yields due to factors such as new breeds and crops, the use of agro chemicals, better production methods and improved management skills. An increasing trend in yields is sometimes offset by drops in yields due to climatic risks or contagious (animal) diseases. Changes in yields can temporarily distort the balance of supply and demand on the market resulting in prices changes. It should be noted that temporary changes in prices might have a longer lasting effect due to adaptations in market supply. These adaptations can affect farm incomes for a longer period.

Several papers describe the volatility of prices on agricultural commodity markets (FAO 2004). Others have studied volatility of yields in combination with prices (Polome *et al.*, 2006). One important issue in these studies is the distribution of these volatilities. In many practical work, the hypothesis of normality is maintained (Just and Weninger, 1999). Another important issue is the existence of an underlying trend in the development of prices and yields. Harmignie *et al.* (2005) show that yearly variations of yields or prices tend to follow either a stochastic process or a cycle, but not a linear trend.

Fluctuations in yields and prices result in strong dynamics in farm incomes. The relationship between yields, prices and farm income is not as straightforward as one might think. It's a complex relationship depending on the cost structure of the farm (direct costs, overhead costs etc.), other agricultural revenues and extra ordinary benefits or costs. The dynamics of farm incomes have been studied in several papers (Phimister *et al.*, 2004; Hergrenes *et al.*, 2001; Mishra and Goodwin, 1997).

Other papers have analysed differences in farm performance. Poppe and Van Meijl (2006) analyse the differences in farm profit, environmental performance and the underlying differences in farm strategy, innovation and management skills. Differences in farm income result in an income distribution with a wide dispersion in each country. Allanson and Hubbard (1999) develop rules to compare income distributions between countries. Fluctuations in incomes and differences in performance between farms can to a large extend be considered as normal fluctuations in farm incomes.

Another line of research focuses on the impact of crises on the economic performance of farms. Mangen and Burrell (2003) for example analyse the different welfare aspect in case of a classical swine fever outbreak. A distinction is made between the different effects of the crises and those who suffer or benefit from these effects.

For proper understanding of the behaviour of farmers and the development of farms it is important to realise that many of the farms combine agriculture with other economic activities (either non-agricultural forms of production on the holding, or employment or self-employment off the farm, or income from financial and real estate assets). Available data on these aspects is however limited. The Farm Structure Survey (FSS) indicated that some 30 percent of EU15 farmers had other activities in 2000. Eurostat's Income of the Agricultural Household Sector (IAHS) statistics suggests that even among households for which farming was the main income source, other income sources provided between a third and a half of the average household income (Eurostat 2002). There is a broad range of academic literature (e.g. Nakajima, 1986) and an accumulating body of empirical evidence that supports the importance of taking a broad view of resource flows when explaining farm-level behaviour (Harrison, 1975; Phimister, 1993; Allanson and Hubbard, 1999; Hegrenes *et al.*, 2001; Findeis, 2002; Offutt, 2002; OECD, 2002, 2004a). Characteristics that influence off-farm employment decisions are studied by Benjamin and Kihmi (2006).

Besides scientific research, income development is a topic described in many monitoring and policy oriented publications at national and European level (Berkhout and Van Bruchem, 2006; EU, 2007). The traditional micro-economic perspective applied in most of these publications will be briefly introduced later.

## 2.3 Methodology and data source

The analyses presented in this chapter are based on FADN data which was established in 1965 (DG-Agri, 2002). The primary aim of the system is to gather accountancy data from farms for the determination of incomes and business analysis of agricultural holdings. It is an important instrument for evaluating income of agricultural holdings and the impacts of the Common Agricultural Policy.

FADN consists of an annual survey carried out by the member states of the European Union. To assure representative data, a sample is constructed based on three criteria: region, economic size and type of farming. Type of farming indicates the most important agricultural activity or set of activities on a farm, for example field crop farms, dairy farms or mixed farms.

The advantage of FADN is that it is a harmonised data source with similar bookkeeping principles in all member states. A further advantage, which is explicitly used in this research, is the micro economic nature of the data source. Detailed information is available on individual holdings, which provide the opportunity to conduct analysis on a holding level and gives insight in the distribution and differences in incomes between holdings. Furthermore it makes it possible to follow the performance of a farm during consecutive years.

The analyses presented in this chapter are based focus on the period 1996 till 2004. 2004 is the first year in which data on the EU-25 is available; some results will be presented on the level and distribution of incomes in the new member states. Based on one year data it is not possible to provide statistics on fluctuations between years. Some more in-depth analyses are conducted for the Netherlands, Spain and Germany. The available dataset can be characterised as an unbalanced panel. Farms participate in the panel for several years. The income definition used in this chapter is the family farm income according to the definition as used by the European Commission.

Section 4 and 5 focus on normal income fluctuations, while section 6 focuses on income crisis in agriculture. In section 4 the traditional micro-economic perspective is described. Based on individual farm observations averages of groups are calculated and presented. This chapter will however show that this traditional approach of comparing group averages over years hides or underestimates fluctuations at farm level because of large differences between farms. Therefore section 5 analyses the developments over years at individual farm level. The heterogeneity of farms and the volatility of incomes at individual farm level are analysed. Heterogeneity is important because there are large differences in the results of individual farms even within the same year or the same type of farming. Heterogeneity is illustrated by showing the range of incomes during a specific year in a specific country. Normal income fluctuations of individual farmers over years are illustrated by a number of measures. The first measure presented is the mean absolute income change from year t-1 to year t. This measure conveys information on the yearly 'shocks' in the income results. Another measure is the coefficient of variation (CV) of farm income on an individual farm. This CV of an individual farm is calculated as the standard deviation of all observation of an individual farm divided by the average of those observations. Furthermore the income distribution is analysed. Farm results in each year are categorised in five quintiles and subsequently it is analysed whether farms stay in the same quintile over years or whether they change their relative position on the income distribution. This sheds light on the issue whether farmers are 'trapped' in a low income situation.

Income crisis is analysed in section 6. Two approaches are used; one is a case based approach in which the impact of specific crisis in the past on the income of farmers is analysed. The second approach estimates the downside risk of farms by simulating the impact of a crisis on the income situation of farms. The shortfall risk is defined as the probability that a farmer will have a negative farm income after the occurrence of an external event.

## 2.4 Structure and development of farm incomes in the EU – the traditional micro-economic perspective

The traditional micro-economic perspective consists of the calculation of group averages and the comparison of averages of groups in different member states, years or types of farming. This section will take this perspective and looks at the changes in the structure of farms and the level of incomes. Farms in the European Union have shown a strong structural development in the last decades. In general, farm size has increased to achieve economies of scale; production has become more specialised with a continuous reduction in labour input and an increase in capital. The restructuring of agriculture has resulted in a concentration of production on a decreased number of farms. Differences in local production circumstances (land quality, weather conditions, water supply) but also economic circumstances (such as cost of production factors, especially land and labour and distance to markets and processors) and socio-economic circumstances (economic viability of rural areas and unemployment), have created or sustained different production structures in different areas in Europe. A clear indicator for these different structures is the (economic) size of farms. Lianos and Parliarou (1986) show that the average farm size over regions in Greek agriculture is found to be related to population pressure (—), industrialisation (+), and mechanisation (+). Figure 2.1 provides a map with average economic size in the FADN regions in Europe. Average farm size is clearly higher in the north-western part of Europe. Due to high labour, land and capital costs there is a high pressure to concentrate production to achieve economies of scale. The average size of farms is also high in the Czech Republic and Slovakia.

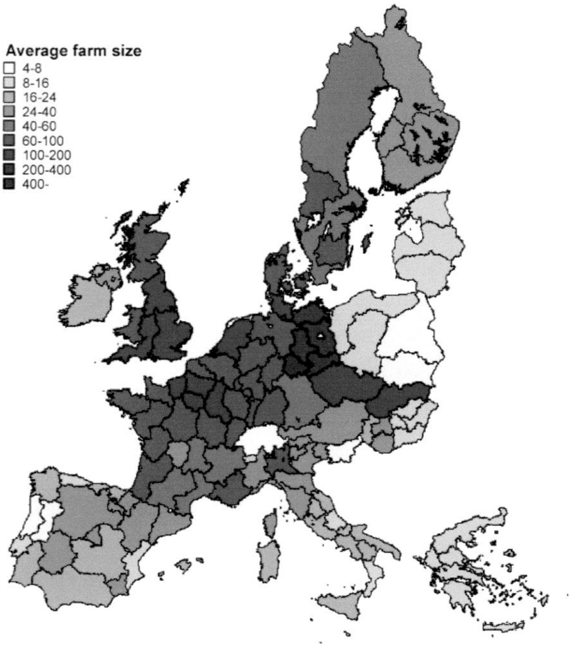


**Figure 2.1.** *Average farm size of farms in EU-25 in ESU (three year average 2002-2004, new member states 2004). © EuroGeographics 2001 for the administrative boundaries. Source: EU-FADN – DG AGRI G-3.*

Figure 2.2 shows the developments in size of farms in relation to changes in farm incomes. Farms in North-western Europe (Germany, France, Belgium, the Netherlands, Luxembourg and Denmark) have increased production more rapidly. In some cases this strategy of growth and improved economies of scale was profitable, but in the Netherlands, France and Denmark the increased size did not compensate lower margins. Mediterranean member states have much smaller farms, but relatively they achieved strong growth in farm size and good improvements in farm income.

Although group averages are frequently presented to illustrate developments in agriculture, they only convey part of the information. Differences in the farm, the farmer and the surroundings cause large differences in the results of farms.

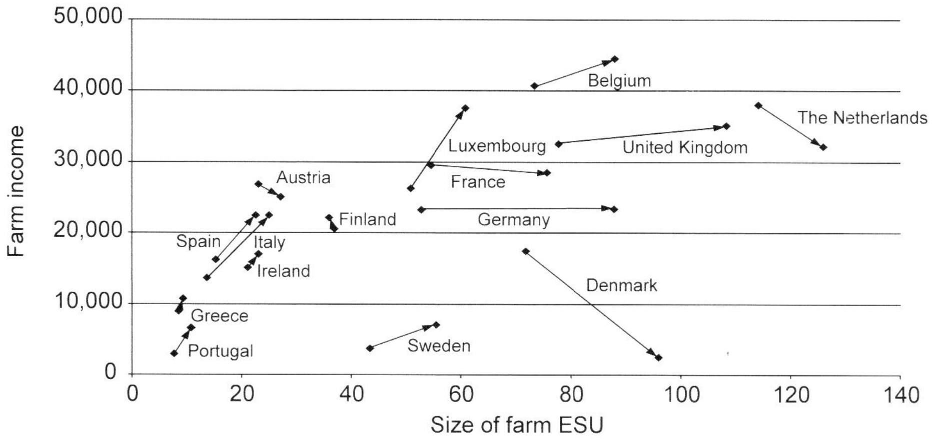


**Figure 2.2.** *Changes in family farm income and farm size (ESU) per EU-15 member state, '1997' to '2003' (three year averages). Source: EU-FADN – DG AGRI G-3.*

## 2.5 Heterogeneity and volatility of farm incomes at individual farm level

This section goes beyond group averages as used in the traditional perspective as presented previously. Group averages are reported in many publications on (the development of) farm incomes. Averages are a good indicator of farm income in case of homogenous groups. There are however many factors that cause large differences between farms. Group averages do not fully show the heterogeneity of farms and the volatility of incomes at farm level.

### *2.5.1 Heterogeneity of incomes within a group of farms*

Heterogeneity is caused by many factors. Differences in the application of modern technologies, size and structure of the farm, environmental and climatic conditions can cause differences in results. Besides these farm characteristics also the characteristics of the farmer, such as training and skills, have a decisive impact on the results (Poppe and Van Meijl 2006).

Figure 2.3 illustrates the large differences between arable farms within member states. The lower end of the bar displays the 5th percentile. The upper end of the bar represents the 95th percentile. The mark within the bar displays the median income. For this specific year and type of farming, the highest heterogeneity with respect to farm incomes are revealed for the Netherlands: 5 percent of the farms achieve incomes higher than 150,000 euro. At the lower end 5 percent of the farms have a negative income of more than 80,000 euro. The median value is just above zero. Other member states with a substantial heterogeneity and a substantial percentage of negative incomes are Germany, the United Kingdom and Sweden. Figure 2.3 is merely an illustration of heterogeneity of farms in the arable sector. The arable sector is not an exception since other types of farming show similar patterns.

### *2.5.2 Volatility of incomes at farm level*

Given the wide range of incomes and the changes in average and median incomes in subsequent years, this raises the question whether farmers collectively move up and down following the trend or that farmers change their relative performance on the income distribution. The latter would mean that farmers are confronted with even higher fluctuations in incomes at farm level. From a farmers perspective the changes in farm income on an individual farm is the most important fluctuation. It reflects the uncertainty a farmer has to cope with. Figure 2.4 gives a clear indication on the large fluctuations at farm level. It displays the mean yearly changes in farm income for different types of farming in Spain, Germany and the Netherlands. The bars display the 25$^{th}$ and 75$^{th}$ percentile and the tick mark displays the median. It means that still 25% of the farms have a mean yearly change that is higher than the top of the bar and 25% have a mean yearly change lower than the bottom of the bar.

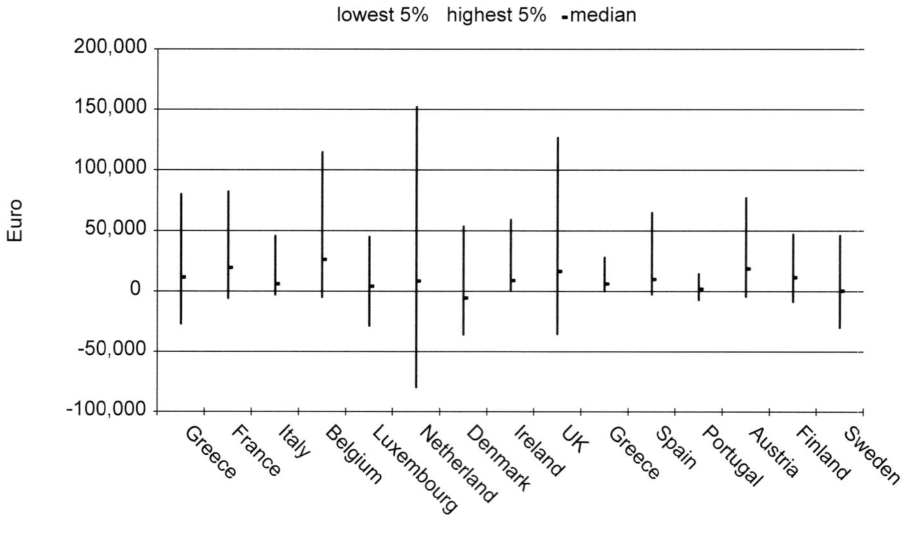


**Figure 2.3.** *Distribution of family farm income on specialised field crop farms EU-15, 2002. Source: EU-FADN – DG AGRI G-3.*

Figure 2.4 shows large differences between types of farming and between member states, but also between farms. The first bar DEU 13 indicates that the median value of yearly changes is 18.000 euro. 25% of these German arable farms have year to year fluctuations of more than 36.000 euro and 25% of the farms have yearly fluctuations of less than 9.000 euro. Similar ranges can be observed for other types of farming in Germany. Dutch farmers in the selected farm types are confronted with the highest yearly changes in farm income. Figure 4 also shows that the largest variety in farm incomes can be found in horticulture and intensive livestock farms. These are the sectors that mainly produce outputs that are not regulated by the CAP. The volatility of the prices is reflected in the volatility of the incomes. The values displayed are absolute values, relative measures in comparison to the level of incomes will be presented in Table 2.2 and subsequent tables and figures.

Comparing the high yearly fluctuations over a period at farm level (Figure 2.4) with the low median incomes in a certain year (Figure 2.3) confirms that trends in median incomes hide important information on the fluctuations that farmers experience, as is partly suggested by the distribution around the median in Figure 2.3. Comparing the yearly fluctuations at farm level as displayed in Figure 2.4 with the yearly changes at sector level (see for example Figure 2.7 for the fluctuation at sector level of field crop farms in the Netherlands) indicates that fluctuations at farm level are substantial higher than at sector level. This implies

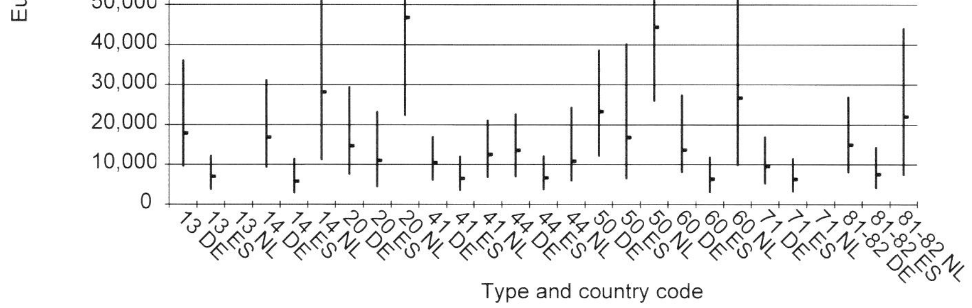


**Figure 2.4.** *Year to year fluctuations of incomes in different types of farming (1996-2004). Type and member state coding: 13- cereals, 14 – field crops, 20 – horticulture, 41-dairy, 50 – intensive livestock, 60/71/81 mixed farms. Source: EU-FADN – DG AGRI G-3.*

that the relative income situation of farmers changes from year to year. Table 2.1 analyses the relative income situation in different member states in different types of farming. One of the underlying questions is whether farms in the lower percentiles of the income distribution are 'trapped' in this low income category, or phrased otherwise do farms in the lowest quintile of incomes in a certain year stay in the lower quintile in the next year, or can they improve their (relative) income position by moving to another quintile. Table 2.1 presents the distribution of specialised dairy farms, intensive livestock farms and mixed farms over quintiles in year t that belonged to a certain quintile in year t-1. It's important to note that the quintiles are based on the relative income position, a farm can change its' relative income position and move to another quintile without any change in absolute income. For example 55.5 percent of the dairy farms in Germany which belong to the lowest quintile in year t-1, still belong to the lowest quintile in year t. 23.5% move up one quintile and 4.5% move up to the highest quintile. Looking at the best performing farms in year t-1 the table shows that almost 40 percent drop back to a less performing quintile, 3.9% of the farms drop back to the lowest quintile. For Spain the percentages for dairy farms are rather similar to those of Germany. Dutch income distribution of dairy farmers is more stable. Roughly 70% of the best performing and worst performing farms in year t-1 still belong to the same quintile in year t.

In all three member states the income distribution of intensive livestock farms is less stable than for dairy farms. For Spain the difference is rather limited in comparison to Germany and the Netherlands. Further analyses have shown that the percentage of farms that stay in the lowest quintile during all the years that they participate in the FADN is very low. For some farms (that have high off-farm income) being trapped in the lowest quintile might be the desired position.

Figure 2.4 presented absolute changes from year to year on a single farm. Table 2.2 presents a EU wide analysis of the differences in volatility for the different types of farming. Volatility is measured as the coefficient of variation of farm income. The coefficient of variation is clearly higher in the intensive livestock sector. In the dairy and grazing livestock sector the incomes are the most stable. Aggregating data over the whole of Europe of course hides large differences between regions. Therefore Figure 2.5 presents the coefficient of variation for the different regions in Europe. The values reflect a combination of factors, such as: the climatic conditions, occurrence of diseases, the type of farming prevalent in the region, crops and animal products produced. The highest volatility of farm incomes can be found in the north-western part of Europe. This contrasts with some previous research and expectations with respect to the climatic circumstances in different parts of Europe. Therefore the right half of Figure 2.5 presents the volatility of total

**Table 2.1**. *Stability of income distribution in different types of farming in different member states (1996-2004).*

| | T | Specialised dairy | | | | | Intensive livestock | | | | | Mixed farms | | | | |
|---|---|---|---|---|---|---|---|---|---|---|---|---|---|---|---|---|
| T-1 | | 1 | 2 | 3 | 4 | 5 | 1 | 2 | 3 | 4 | 5 | 1 | 2 | 3 | 4 | 5 |
| Germany | 1 | 55.5 | 23.5 | 10.3 | 6.3 | 4.5 | 46.1 | 23.6 | 14.0 | 9.7 | 6.5 | 46.5 | 26.6 | 12.2 | 10.0 | 4.7 |
| | 2 | 25.5 | 36.9 | 23.3 | 10.7 | 3.5 | 26.8 | 33.1 | 24.0 | 10.3 | 5.8 | 25.7 | 32.3 | 23.6 | 11.9 | 6.4 |
| | 3 | 10.0 | 25.4 | 33.7 | 22.5 | 8.4 | 10.5 | 32.4 | 27.6 | 20.4 | 9.1 | 15.0 | 23.4 | 31.4 | 21.2 | 8.9 |
| | 4 | 5.6 | 11.5 | 25.7 | 36.5 | 20.6 | 7.5 | 11.5 | 24.5 | 30.1 | 26.5 | 6.8 | 12.1 | 24.4 | 34.6 | 22.1 |
| | 5 | 3.9 | 4.5 | 8.7 | 22.9 | 60.0 | 3.5 | 4.9 | 10.2 | 28.4 | 52.9 | 5.2 | 4.9 | 8.8 | 23.4 | 57.6 |
| Netherlands | 1 | 69.1 | 22.3 | 5.2 | 2.4 | 0.9 | 48.6 | 21.4 | 14.8 | 9.6 | 5.6 | 44.2 | 30.0 | 13.4 | 8.3 | 4.0 |
| | 2 | 25.0 | 48.8 | 20.0 | 5.5 | 0.7 | 21.3 | 35.1 | 21.6 | 13.1 | 8.9 | 19.3 | 28.8 | 40.7 | 7.1 | 4.2 |
| | 3 | 7.0 | 23.1 | 48.3 | 18.8 | 2.8 | 9.9 | 27.9 | 29.6 | 24.2 | 8.4 | 11.1 | 17.1 | 31.1 | 29.6 | 11.2 |
| | 4 | 3.3 | 6.1 | 20.8 | 52.6 | 17.2 | 9.0 | 18.4 | 24.4 | 34.4 | 13.8 | 9.9 | 22.6 | 27.9 | 29.9 | 9.8 |
| | 5 | 1.3 | 1.1 | 3.8 | 19.0 | 74.8 | 9.8 | 3.0 | 11.8 | 15.9 | 59.5 | 9.0 | 3.6 | 5.8 | 9.2 | 72.4 |
| Spain | 1 | 48.6 | 29.7 | 11.9 | 6.4 | 3.3 | 48.4 | 20.4 | 13.3 | 9.4 | 8.5 | 47.8 | 15.1 | 30.5 | 3.5 | 3.1 |
| | 2 | 29.7 | 34.6 | 24.0 | 9.3 | 2.4 | 21.7 | 45.5 | 19.0 | 7.3 | 6.4 | 23.8 | 41.5 | 24.5 | 8.9 | 1.4 |
| | 3 | 16.2 | 23.0 | 33.3 | 21.6 | 5.9 | 16.7 | 18.0 | 36.1 | 19.9 | 9.3 | 12.0 | 25.8 | 34.2 | 23.0 | 5.0 |
| | 4 | 7.2 | 10.4 | 23.7 | 39.9 | 18.9 | 7.1 | 12.2 | 26.6 | 35.4 | 18.7 | 4.9 | 6.7 | 21.2 | 48.4 | 18.8 |
| | 5 | 3.9 | 3.0 | 7.2 | 22.0 | 63.9 | 8.3 | 2.4 | 15.2 | 26.0 | 48.2 | 2.1 | 2.4 | 4.9 | 25.6 | 65.0 |

Source: EU-FADN – DG AGRI G-3.

**Table 2.2.** *Volatility at farm level EU-15 (1996-2004).*

| Type of farming | Coefficient of variation of Family Farm Income Tukey M-estimator |
|---|---|
| Field crops | 0.31 |
| Horticulture | 0.37 |
| Wine | 0.33 |
| Other permanent crops | 0.33 |
| Milk | 0.28 |
| Grazing livestock | 0.31 |
| Intensive livestock | 0.53 |
| Mixed | 0.29 |

Source: EU-FADN – DG AGRI G-3.

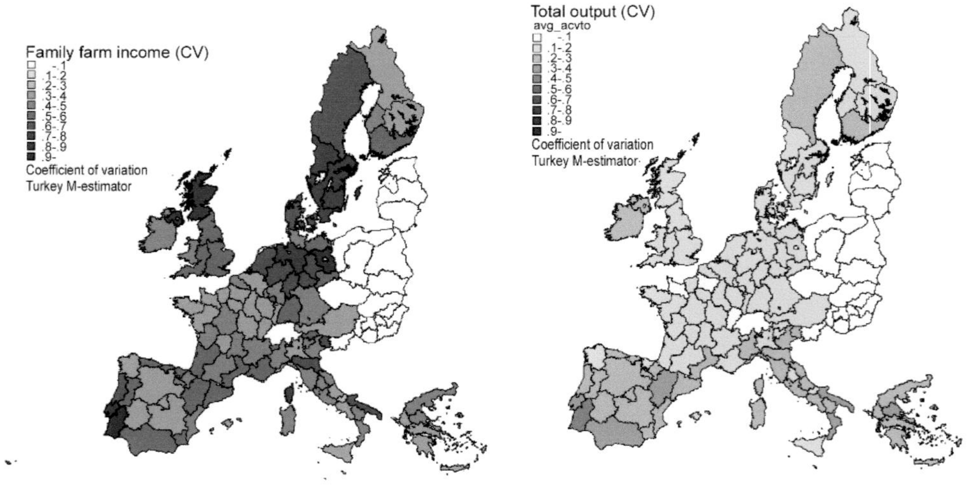


**Figure 2.5.** *Within-farm volatility of family farm income total output in EU-15. © EuroGeographics 2001 for the administrative boundaries. Source: EU-FADN – DG AGRI G-3.*

output. This pictures shows that the output volatility, which is conceptually more correlated with the production circumstances, is higher in the southern European countries and to a lesser extent in the Nordic countries. Comparison of both halves of the figure reveals that the volatilities of farm income are much higher than those of production value. Farm income is much more volatile because it is a residual indicator. More specific analysis have shown that although the production volatility in Spain is higher than in Germany or the Netherlands, the volatility in net value added or family farm income is higher in the latter. This is mainly caused by the differences in the (financial) structure of farms.

Table 2.3 gives an overview of differences in farm volatility in different member states in the EU and within different types of farming. The values display the index compared to the overall central tendency of the coefficient of variation (0.3144). In most member states the intensive livestock sector has the highest within-farm volatility. Exceptions are Austria, Portugal and Finland where the other permanent crops sector (fruit) shows the highest volatility. The numbers for Italy are quite different, the intensive livestock sector but also all other sectors show a low volatility. The volatility of incomes of mixed farms is rather high in some member states. Although diversification can be used as a risk management strategy, the overall volatility strongly depends on the agricultural activities on the mixed farms. In Germany, Denmark and the Netherlands, it is for example rather common that mixed farms produce pigs. The volatility of the revenues from the pig production

**Table 2.3.** *Index of within-farm income volatility in different member states and types of farming 1996-2004 (index = 100 equals 0.3144).*

| | Field-crops | Horti-culture | Wine | Other permanent crops | Dairy | Grazing livestock | Intensive livestock | Mixed |
|---|---|---|---|---|---|---|---|---|
| BE | 87 | 101 | | 145 | 61 | 81 | 193 | 94 |
| DK | 109 | 158 | | 177 | 102 | | 264 | 118 |
| DE | 134 | 130 | 107 | 144 | 101 | 121 | 191 | 158 |
| EL | 102 | 107 | 75 | 109 | 95 | 68 | 204 | 59 |
| ES | 97 | 145 | 108 | 122 | 99 | 90 | 153 | 76 |
| FR | 113 | 126 | 145 | 182 | 85 | 99 | 179 | 96 |
| IE | 82 | | | | 73 | 121 | 200 | 84 |
| IT | 78 | 82 | 78 | 85 | 75 | 75 | 85 | 61 |
| LU | 181 | | 121 | | 96 | 98 | 340 | 180 |
| NL | 217 | 174 | | 146 | 113 | 291 | 475 | 216 |
| AT | 89 | 94 | 117 | 154 | 82 | 79 | 127 | 98 |
| PT | 116 | 111 | 171 | 184 | 112 | 112 | 169 | 126 |
| FI | 154 | 181 | | 222 | 76 | 101 | 139 | 105 |
| SE | 186 | | | | 177 | 140 | 231 | 148 |
| UK | 137 | 89 | | 98 | 109 | 171 | 173 | 131 |

Source: EU-FADN – DG AGRI G-3.

has a strong impact on the total farm income volatility. Whether diversified farms have a lower volatility in comparison to specialised farms strongly depends on the types of activities on those farms.

Table 2.3 also shows large differences in the volatility in the field crops sector. The Netherlands has the highest volatility followed by Sweden and Luxemburg. The coefficient of variation in the field crops sector is further specified in Figure 2.6. Besides climatic conditions the differences are explained by the differences in cropping patterns. For example in the Netherlands potatoes and onions are important products. These products (except for starch potatoes) are hardly regulated by the CAP. Therefore changes in yields, due to heavy rainfall, crop diseases, draught or other conditions, have a strong impact on the price level and on the revenues and profits of farms. Figure 2.6 clearly shows that there can be large differences within a member state, and even within one specific type of farming. Besides the cropping pattern, also the (financial) structure of the farm is important. Although the arable farms in southern Europe have more unpredictable yields due to climatic conditions and especially drought, this figure shows that farm incomes in northern Europe are more volatile.

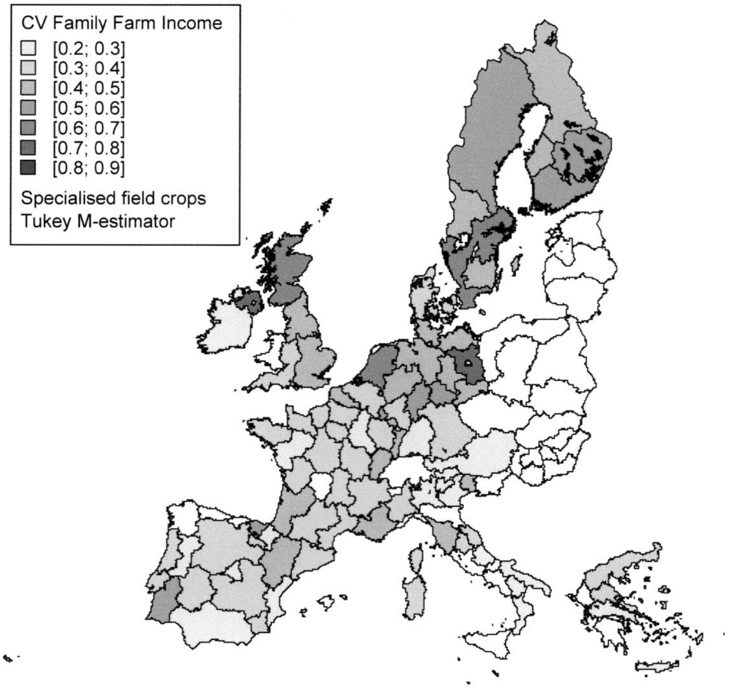


**Figure 2.6.** *Within-farm volatility of family farm income on specialised field crop farms in EU15. © EuroGeographics 2001 for the administrative boundaries. Source: EU-FADN – DG AGRI G-3.*

The previous analyses have shown that farms are confronted with volatile production outputs and volatile incomes. This is an important aspect to consider in analysing incomes, making inferences about the developments of incomes and in policy evaluations. Another important conclusion directly affecting this research is that normal business fluctuations already cause a large volatility in farm incomes. Based on FADN data it is difficult to separate normal business fluctuations from crises. Some farms can end up in a financial crises due to normal business fluctuations, other farms are able to survive an external crisis.

## 2.6 Impact of crises on farm incomes

FADN, being a bookkeeping system, does not contain indicators whether certain special events occurred on the farm during a year[1]. To analyse the impact of external events on results of a farm it's therefore necessary to either deduce from the data that something special occurred or by using other information / knowledge on the occurrence and scope of external events. The application of the first approach is complicated by the fact that bookkeeping data are a summation of all effects over a period of time (business decisions, external effects, farm risks, personal risks etc.). Therefore, the latter approach is used. In several case studies the impact of a crisis on farm results, the distribution of results and supplemental effects are analysed. The second approach looks at the shortfall risk. The shortfall risk is defined as the probability that a farmer will have a negative farm income after the occurrence of an external event.

## 2.7 Case descriptions: impact of crises on farm incomes

To provide more in-depth analyses of the impact of external crises a number of cases are analysed. The cases clearly show the usefulness of FADN data in analyzing time series of farm incomes. However it often requires local knowledge to assure a good interpretation of the data. Two cases will be illustrated based on Figure 2.7.

In early 1997 pig farms, mainly in the southern part of the Netherlands, were infected by classical swine fever. More than 10% of the specialised pig farms were confronted with stamping out of the animals. The negative impact on income per pig farmer was around 45,000 Euro on farms with stamping out after compensation by the government. During the swine fever period most pig farmers, outside the

[1] During the submission of data to Brussels, checks on the quality are performed. A crises would in many cases result in a warning because results are different from expected. The national FADN liaison is able to confirm the supplied data and supply the reason for these deviating values. Although this information could contain information on the type of crisis that occurred, the information is not collected and stored in a systematic way. Therefore the use of this additional information is considered to be very complicated.

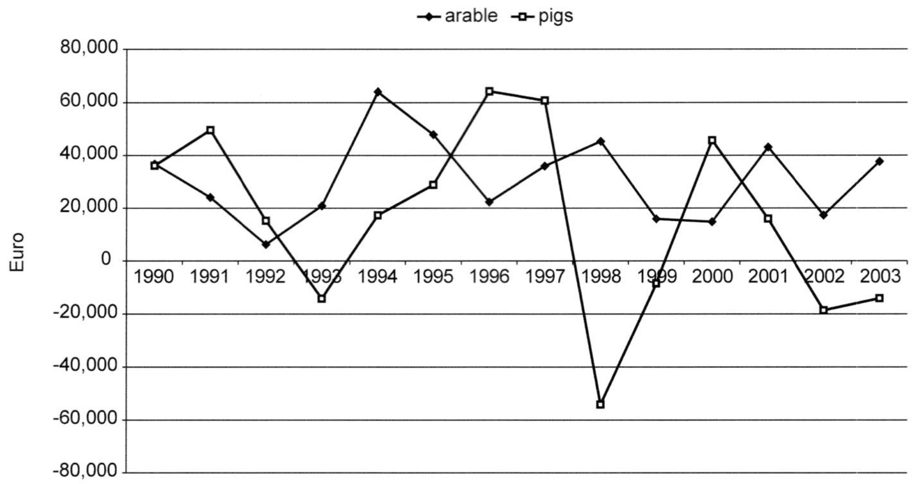


**Figure 2.7.** *Fluctuations in average family farm income of arable and pig farms in the Netherlands. Source: EU-FADN – DG AGRI G-3.*

region, had still a rather high income. This was caused by the high level of prices during that period. Partly this was the consequence of the reduced production volume in the Netherlands, an exporting country with a market share in the EU of ten percent at that time. During the swine fever period, 2 million pigs and piglets were destroyed, 15% of the stock in the Netherlands. The higher price level stimulated production in other EU member states during 1997 and 1998. Prices went down in 1998 as a consequence of the above normal supply on the market. In November 1998 prices came at the lowest levels after the Second World War. This had very clear effects on the level of incomes during that year.

Around 60% of arable farmers in the Netherlands were confronted in the autumn of 1998 (September, October) with too much rain to harvest their crops, mainly potatoes as well as onions. The amount of rainfall in these months was the most severe in the 20th Century. Under these extreme condition prices of ware potatoes became 100% higher during the marketing year 1998 than under normal climatic condition. Due to these high prices, the returns of ware and plant potatoes were on average 40% per farm higher than in normal years. The average results of arable farms in 1998/1999 improved for a large part as a consequence of these higher prices and by the compensation for harvest losses. The average family farm income on arable farms was clearly higher than in the previous two years and much higher than in the next two years (1999 and 2000). About 35% of the arable farmers still had a lower income than under normal conditions. Partly these were producers of starch potatoes, who didn't benefit from higher prices due to fact that starch potatoes prices are regulated by the CAP.

## 2.8 Downside risk and farm incomes

The possibilities of farms to cope with the occurrence of external events differ strongly. Shortfall risk will be specified as the percentage of farms in a region or in a country that will have a farm income of less than zero due to a price or revenue decrease as a consequence of a possible crisis. A distinction is made between including and excluding opportunity costs. Cost of own labour is calculated as the average of paid labour in a specific region (Niemi and Ahlstedt, 2007), cost of own assets is calculated as 4% of own equity.

The analysis focused on specialised farms (specialised pig and arable farms) that were in the sample for the three succeeding years 2002-2004. For every farm, the normal uncertainty in the revenues was calculated. Based on the financial structure of the farm an analysis was made how robust a farm would be to survive an external event that would reduce the output value with 30 percent. In order to show the robustness of the farm itself, the assumption was made that there are no indemnity payments and that the external event does not change the cost structure of the farm. To establish the financial robustness of farm, farms were categorised in five categories:

- Family farm income higher than opportunity costs.
- Family farm income is still positive after the external event.
- Family farm is negative, but postponing redemption (assumption: redemption equals depreciation ) is an option.
- Family farm income cannot be compensated with postponing redemption. Unless the farmer has liquidities to compensate for the negative income, financial distress will be the result.
- Family farm income is already negative before external shock; the external event only deteriorates the situation.

The results of the simulations are displayed in the following graphs (Figure 2.8).

The pie charts present large differences between member states. For arable farmers, the charts show that a large percentage of farmers in Spain (90%) can cope with a 30% reduction in output (still a positive income). In Germany this percentage is 32%, in Hungary 37% and in the Netherlands 18%. The situation is worse in the intensive livestock sector. In Spain 49% of the farms will still have a positive income after a 30% shock. In Germany this percentage is 4% and in Hungary 13%. In the Netherlands this percentage is close to zero, but this figure is a partly biased

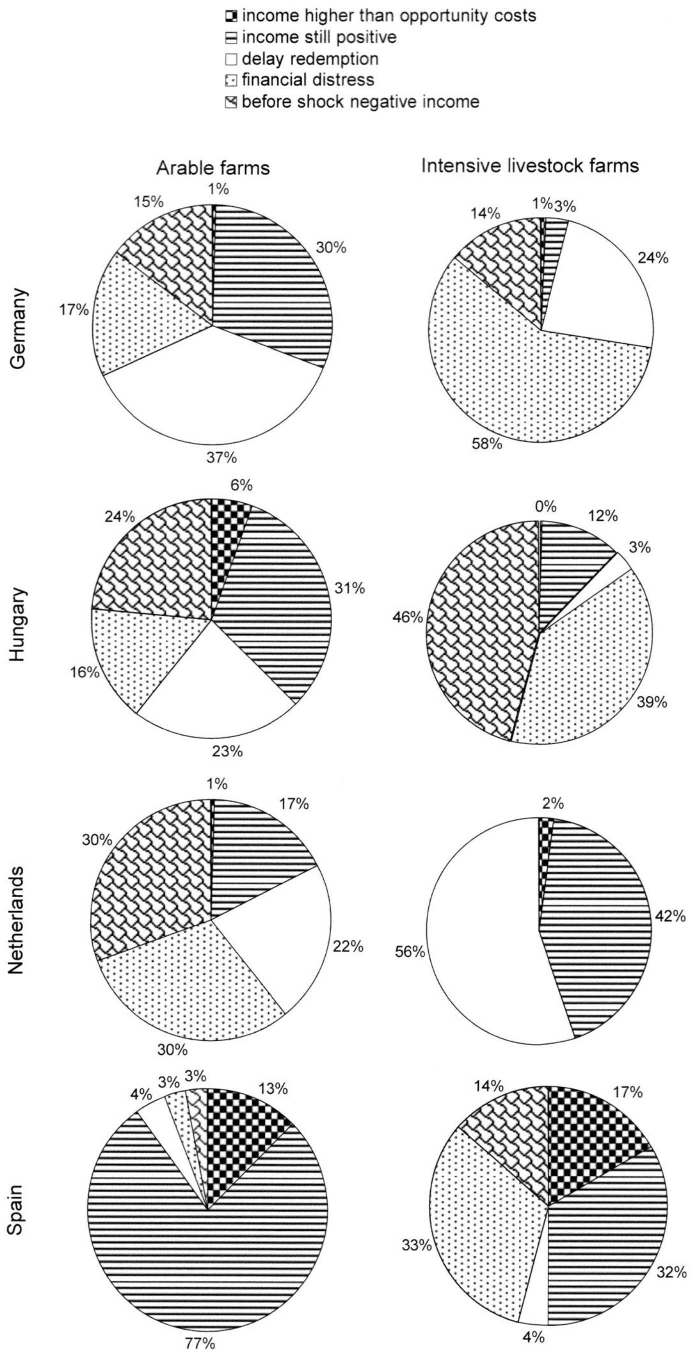


**Figure 2.8.** *Financial robustness of farms after external crises, 30% drop in output revenue (first column of graphs are specialised arable farms, second column specialised pig farms). Source: EU-FADN – DG AGRI G-3; Hungarian data from AKI.*

due to the low income years 2002 and 2003, which had a negative impact on the starting situation (already 56% of the farms had a negative average income during the period 2002-2004). Whether financial distress leads to the bankruptcy of the farm depends on many other factors such as farm wealth, off-farm wealth, off-farm income. These graphs show that although agriculture in Southern Europe is more vulnerable to draught, the financial risks are larger in North Western Europe due to small margins. So, although the climatic conditions have an impact on the volatility of production, the volatility of farm incomes is strongly affected by the (financial) structure of the farm. These structures are also dependent on the risks that farmers have learnt to cope with. Further analyses (not reported here) have shown that there is no strong link between the size of the farm and the extent to which a farm can cope with an external crisis.

## 2.9 Discussion and conclusions

### *2.9.1 Conclusions on income volatility and income crisis*

Farmers are confronted with a wide range of factors that affect their income. Besides a continues increase in productivity, fluctuations in yields due to climatic conditions and fluctuations of prices of outputs and inputs strongly affect the levels of incomes. Contagious diseases affecting the production of crops and animals are external events that can cause a crisis on a (group of) farm(s).

The analyses of individual farm data show strong fluctuations in farm income. Large differences between different member states, regions and sectors occur. Furthermore, the quantitative analyses show that there are strong differences between farms within the same type of farming. Size of farm only provides a small explanation of these differences. Average farm incomes only convey a limited amount of information. On the one hand they do not show that even with a positive average there can be a large group of farms with low or even negative incomes. On the other hand the strong fluctuations of incomes and the strong changes in the relative income position of farms stress the importance to look at a long year average to draw meaningful conclusions over the level of income and the standard of living of individual farmers.

The availability of information on off-farm income is still limited, especially in the FADN framework, but there are clear indications that the importance of off-farm income is increasing. Off-farm income is more stable than farm income and thus provides a cushion for farm income fluctuations. Also off-farm assets are essential in understanding farm behavior and their ability to cope with crisis. Therefore it is not possible to predict with current datasets whether a crisis will possible lead to bankruptcy or whether sufficient resources are available to absorb a shock.

The analyses also show that there are large differences in the shortfall risk of farms. Simulated crises show that farms in North Western Europe have a much higher shortfall risk due to the structure of farming. Small margins make it much more difficult to absorb a shock in the short term. The case descriptions show the market response to certain events increases the fluctuations. The market response after the swine fever outbreak led to a strong increase in production due to temporary higher prices (due to a reduced supply) and therefore to a collapse of prices and farm incomes.

Although FADN data can clearly quantify the impact of factors such as heavy rainfall and classical swine fever on farm incomes, a clear distinction between normal volatility and income crises due to an external event is difficult to make. FADN data is an aggregation of all events that happened during a year on the farm. Normal fluctuations are substantial and if a adverse outcomes are observed it requires further information to establish whether this was due to external crises such as heavy rainfall, or that it was due to other circumstances such as bad management, or illness of the farmer.

### 2.9.2 Discussion on the use of FADN data

If the European Union moves towards systems of income stabilisation and (crisis) risk management the question arises if the FADN could play a role to monitor stabilisation programs. Our analysis shows that the FADN could be beneficial in a number of ways.

FADN is a useful and established tool to monitor income and situations of low farm income; however it is criticised of not providing information on non-farm income and household income (Court of Auditors, 2002). Due to the lack of information on non farm resources it is also not easy to predict what will happen on a farm after a crisis.

FADN is a good tool, as our analysis showed, to monitor normal business risk and to assess the effect of event driven crisis risk on the viability of farms. It also makes it possible to monitor if stabilisation-programs are effective; FADN is a good tool to check the payments of national and regional authorities in relation to regulations on state aid.

However the FADN in its current state seems not a perfect tool to assess the need for crisis-risk management actions by governments. There are a number of reasons for this.

The analysis in this report shows large differences in business risk and large variability in year-to-year income at farm level. This means that a low income or a large drop in income, is often not the effect of an event-driven crisis-risk. In other words: an event-driven income crisis will lead to a drop in income but many drops in income are not due to event-driven crisis risks. To improve the usefulness of the FADN for this type of data analyses it is suggested to add a variable in the data per farm per year to record if a farm experienced a crisis-event (with a list of pre-defined codes).

The main problems of event-driven crises are often not the direct effects (e.g. stamping out animals in a contagious disease like classical swine fever) but the effects due to the market response that changes the business cycle (Mangen and Burell, 2003). It are these effects that will especially show up in FADN data.

Several event-driven crises affect only a small number of farmers (sometimes this is even true for crises with large effects in market response) and this will not be revealed in FADN as it is not representative for small samples. Examples include avian influenza in the Netherlands and heavy snowfall in a tree-nursery area in the Netherlands.

It takes on average about 2 years before FADN data are available at EU-27 level in Brussels. This is far too late, if it has to trigger actions of authorities to deal with a crisis. However it can be used as a check or for ex-post analysis.

In case income-stabilisation programs would be targeted on yield or revenue (yield x price) risk, production (harvest) and price statistics would be more beneficial than FADN data. These are more timely available and are based on standard definitions. In the FADN data yields for many products are not or not very well recorded. Partly this is due to the high (and increasing) heterogeneity of products and partly due to the fact that in many member states data are obtained from bookkeeping data for tax purposes (where yield information is not very important).

# Chapter 3

# Risk of low incomes under different policy scenarios

*Edward Majewski, Adam Was, Waldemar Guba, Graham Dalton and Joanna Landmesser*

## 3.1 Objectives

The main objective was to assess the risk of farmers in the European Union and to analyse the impact of agricultural policy changes on the main components of income namely price and production risks. In order to achieve this, qualitative considerations and quantitative analyses covering the period 2004 – 2018 have been made. Future policy scenarios have been defined, taking into account likely Common Agricultural Policy developments, including possible outcomes of the Doha round of the WTO negotiations. Results of the qualitative analysis have been expressed as quantitative values which were used in simulations of farm incomes with the use of a farm level Monte-Carlo simulation model.

## 3.2 Materials and methods

The general approach was to compare the potential risk of low incomes in future periods (2013, 2108), with the situation in the year 2004 taken as a reference. The basis for the qualitative part of the analyses was FADN data for the years 2002-2004. Also some historic statistical data were used, mainly on prices and yields, in order to describe trends which supported predictions of basic parameters. Simulations were made for a set of policy scenarios consisting of two extreme scenarios (liberal and protectionist) and three likely policy options, differentiated by a degree of liberalisation of the EU Common Agricultural Policy (CAP).

### *3.2.1 Policy scenarios*

Traditional CAP-based market price support measures have played an important role in reducing price risk in EU agriculture. The successive reforms (1992, 1999, 2003) have gradually turned more of this support to direct payments, which from 2006 onwards are mostly decoupled. The provisions of the forthcoming new commitments within WTO may force the EU to a further liberalisation of market policy (Swinbank, 2005a).

Enhanced market access and lower internal prices strengthen the links with world market developments and, in general, (have and will) increase price risks in EU farming. On the other hand, decoupled direct payments increasingly insulate farmer's incomes from production and price variability. Further decreases in institutional prices, limitations on intervention purchases and enhanced external competition may radically increase the exposure of EU farmers to price risks. Even though world market prices may increase and become less variable as a result of farm and trade policy reform, e.g. due to WTO agreement, the volatility of EU prices is expected to be greater than in a more protective policy environment.

The WTO commitments mainly sets constraints on the form of CAP support. While the magnitude of total support being delivered to EU agriculture has not changed substantially since 1990, the forms of the support have evolved significantly (OECD, 2005). The CAP reform of 2003 which decoupled most direct payments turning them into green box category is deemed to anticipate many of the new targets of the Doha Round. Nevertheless, the ultimate outcomes of the Doha round may potentially put new pressures on the CAP. In particular the withdrawal of export subsidies and provisions within the market access pillar may be conducive to new substantial rearrangements in the CAP, affecting some CMOs more than others (see e.g. Swinbank, 2005b and numerous reports in Agra Europe Weekly (various issues).

Another driving force influencing future policy choices will certainly be accelerated by debates about the size and allocation of the EU budget as a result of changing public expectations and increasing scrutiny as regards the role and the efficiency of the CAP. As far as the EU budget debate is concerned the recent negotiations on the new financial perspective 2007-2013 clearly show how difficult it is in the enlarged EU and in the context of new global challenges to reach compromises on the EU farm budget which still accounts for more than 1/3 of total EU expenditures. This situation appears to have little consequence for the CAP up to 2013, however, it may have significant implications for the CAP beyond 2013 as public expectations towards the CAP are changing especially for production support. In the future also decoupled income support may no longer find sufficient public legitimacy without linking it even more closely to the provision of public goods desired by society.

In conclusion the trend of long-term agricultural policy shows a gradual liberalisation, which very likely will continue in the future. This assumption was a foundation for the policy scenarios formulation (Table 3.1). The **Base scenario** represents the historic reference reflecting the present policies and market conditions in years 2002-2004. The **Most likely scenario** reflects the policies and market situation as expected in 2013. It assumes a continuation of existing (2006) policies, with some minor changes, including 10% modulation of direct payments. For the year 2018 there were two scenarios constructed, which refer to the probable CAP evolution mainly induced by a prospective new WTO deal – **Likely A** and **Likely B**, both include further liberalisation of market policies, full decoupling of direct support and shifts of budgetary funds out of pillar 1 (so called modulation). They differ in terms of the degree of liberalisation: the **Likely B** assumes a greater reduction in market price support and in the direct support (20% mandatory modulation instead of 10% in **Likely A** and a ceiling of 100,000 euro of direct payments per farm, compared with no such limits in **Likely A**). In addition to the likely scenarios two extreme scenarios for 2018 – **Liberal** (a complete removal of subsidies for agricultural sector) and **Protectionist** (return to the Agenda 2000 type of policy) were also created for comparison.

**Table 3.1.** *Policy scenarios selected for risk simulation.*

| | | |
|---|---|---|
| 2004 | Base | A historic reference |
| 2013 | Most Likely | Most Likely; Existing policy continued (with minor changes), mandatory modulation – 10% |
| 2018 | Likely A | Lower level of market price support, full de-coupling*, mandatory modulation – 10% |
| | Likely B | Further decrease in market price support; full de-coupling*, ceiling – 100,000 euro, mandatory modulation – 20% |
| | Liberal | Market intervention measures removed, no direct payments |
| | Protectionist | Return to 'pre-CAP reform' type of policy – stronger market protection |

* ESP, NL – historical model of SPS. D, PL, H – regional model of SPS.

### 3.2.2 Member states

Five member states have been included in the analysis: Germany, Hungary, Netherlands, Poland and Spain. This set of member states reflects the diversity of EU farming with respect to the risk implications of policy changes. Agricultural sectors in these member states represent different farm structures as well as productivity and technological excellence levels, various climatic conditions, and thus production structure. Both 'new' and 'established' member states are represented, with different degrees of experience with the CAP environment.

### 3.2.3 Simulation model

The risk of low incomes was estimated using the Monte Carlo stochastic simulation method by constructing farm level models (@Risk). For each farm type mean levels and distributions of farm incomes were estimated. The key variables of the model were as follows:

1. Structural – describing farm types (size of activities, and average yields, prices, inputs and costs).
2. Standard deviation as a volatility measure for yields, commodity prices and selected prices (costs) of variable inputs (fixed costs were introduced as constant values specific for each farm type).
3. Cross correlations (input-output, input-input) and market (price-price, price-yield; yield-yield).

For simplification a normal distribution for all variables was assumed. The distributions were truncated with respect to the downside tail to prevent simulating negative yields, whilst for prices this value was, optionally, $\bar{x} - 2\sigma$ or 0 or the intervention price, depending on which was the highest.

Farm incomes and profits for each farm type simulated in the model were calculated in accordance with FADN income definitions.

### 3.2.4 Farm types

Farm types selected for modelling have been clustered according to their production type and economic size. In selecting farm types several criteria such as policy relevance, importance for member states considered in the analysis (based on distribution of FADN farm types in the EU) and expected vulnerability to policy change were taken into account. Consequently, the following farm types based on FADN typology were selected:

A. Common for all 5 member states involved:

Farm type 13 – specialist cereals, oilseed and protein crops
Farm type 41 – specialist dairying
Farm type 50 – specialist granivores
Farm type 81-82 – field crops-grazing livestock combined and various crops and livestock combined

B. Specific for some of the member states:
Farm type 14 – general cropping (Germany, Netherlands)
Farm type 60 – mixed cropping (Hungary, Poland)
Farm type 71 – mixed livestock, mainly grazing livestock (Poland)

It was decided to select farms with at least 8 ESU since smaller farms have less and less significance as market players and they increasingly depend on other income sources. Although the number of such farms is very high in some member states (e.g. Poland), their share in the use of agricultural land, marketable production as well as their contribution to total ESU is very low. Consequently, the following size clusters described in ESU were taken for analysis: 8-16 ESU, 16-40 ESU, 40-100 ESU, >100 ESU. The final number of farm types modelled varies in each member state (Table 3.2) depending on the significance of the farm production types and the existence of farm size clusters.

**Table 3.2.** *Farm types considered, estimated number of farm models for simulation and share of farms of specific type in total number of farms above 2 ESU.*

| Member state/Farm type | Germany | Hungary | Netherlands | Poland | Spain |
|---|---|---|---|---|---|
| 13 | 17% | 28% | - | 3% | 13% |
| 14 | 8% | - | 13% | - | - |
| 41 | 24% | 4% | 26% | 6% | 4% |
| 50 | 3% | 4% | 7% | 7% | 3% |
| 81-82 | 15% | 11% | 5% | 23% | 3% |
| 60 | - | 12% | - | 10% | - |
| 71 | - | - | - | 14% | - |
| Share of total farms represented in the sample | 67% | 59% | 51% | 63% | 23% |
| Total number of models | 108 | 114 | 72 | 120 | 96 |

### 3.2.5 Data sources for model calibration

Model parameters for the base model were derived mainly from FADN. Structural parameters introduced to the model were calculated as mean values for the years 2002-2004. FADN data as well as general statistics for the years 1996-2004 were also used in order to estimate standard deviations for the key model parameters and correlations.

For Poland additional data for the period 1997–2001 from the Farm Survey were merged with the FADN results to obtain a longer time series required for volatility estimation for the variables used in the model[2].

### 3.2.6 Price assumptions for policy scenarios

The impact EU policy changes (level of liberalisation or protection) on the **price level** of EU commodities was established as follows:

1. Price projections for the EU and World market for the years 2007-2016 taken from the *OECD-FAO (2007) Agricultural Outlook 2007-2016*[3].
2. A measure of the price gap between the EU and the World market (see Annex Table B.2) was based on estimates of nominal protection using coefficients for EU farm prices in the PSE calculations in the *OECD (2007a) Monitoring and Evaluation* report.
3. The impact of liberalisation of farm and trade policies world wide on the level of World market prices was based on the assumptions in the reports of the *OECD (2007b)* and *FAPRI (2005)* (see Annex Figures B.1 and B.2).

It was assumed that with no change in the CAP internal EU prices would develop in line with the price projections for the EU market presented by OECD-FAO (2007) Agricultural Outlook 2007-2016. Partial liberalisation of market and trade policy would in the long term bring EU prices closer to world market prices, but the scope of price changes would depend on the commodity and the initial distance between the respective prices[4] (see Table 3.3 and annex Table B.1). In

[2] Since in new member states FADNs have been established very recently they provide data for the years 2002-2004 only. For the Polish model the additional source of data was the Farm Survey conducted by the Institute of Agricultural and Food Economics. Because representation of farms in the Farm Survey is much smaller (about 1000 farms) compared with FADN (12,000 farms in the year 2004) a merged data base was created of all farms from the Survey which represent farm types selected for simulations, with a randomly drawn 10% of the FADN farms population from respective farm types.

[3] The price scenarios refer to nominal price levels. The year 2016 prices (ending year of the recent projection) were used to represent the situation in 2018.

[4] Elements of the market and trade liberalisation, such as changes in tariffs, tariff quotas and direct intervention have not been considered explicitly. Market policy scenarios are described directly by their price effects.

**Table 3.3.** *EU Price change indices for policy scenarios (2005 = 100).*

| | OECD-FAO (2007) Projection | | Assumed price change indices (2005 = 100) | | | | |
|---|---|---|---|---|---|---|---|
| | Price 2005 (EUR) | Price 2013 (2005=100) | Most likely 2018 | Likely A 2018 - higher protection | Likely B 2018 - lower protection | Liberal 2018 | Protectionist 2018 |
| Wheat | 118.35 | 99 | 99 | 99 | 99 | 94 | 109 |
| Coarse grains (barley) | 104.35 | 102 | 102 | 101 | 101 | 96 | 111 |
| Corn | 130.61 | 95 | 95 | 93 | 93 | 89 | 103 |
| Oilseed | 251.63 | 99 | 99 | 100 | 100 | 95 | 110 |
| Potatoes | 115 | 100 | 100 | 100 | 100 | 95 | 105 |
| Sugar beet | 46.72 | 56 | 56 | 56 | 47 | 43 | 100 |
| Milk | 29.69 | 86 | 86 | 87 | 78 | 74 | 104 |
| Beef | 243.20 | 106 | 106 | 107 | 91 | 70 | 118 |
| Pork | 127.44 | 108 | 108 | 112 | 106 | 101 | 117 |
| Poultry | 101.54 | 107 | 107 | 110 | 93 | 77 | 115 |

the Liberal (2018) scenario full alignment of EU prices with world market prices is expected, whilst in the Protectionist scenario it is assumed that prices for sugar and milk 'return' to the pre-reform level and for other commodities increase above the baseline. The projections served to calculate price indices for 2013 and 2018 scenarios (see Annex Figure B.3 and B.4).

Assumptions on the impact of policy scenarios on **price variability** have been made on the basis of a historical comparative analysis of price volatility in the EU and world markets. The analysis confirmed the expectation of much lower price volatility in the EU market compared with the world market level. In general, the differences in price volatility (measured by standard deviations) reflect the impact of market price support and the restrictiveness of market intervention instruments to price transmission in each sector as well as specific structural conditions. The greatest differences were detected for sugar (world market volatility = 3 times EU market volatility), dairy (2 times) and wheat (1.5) and lowest for pork (1.2), cattle (1.2) and poultry. The analysis of the price volatility in the new member states has been conducted on the base of time series limited to the 1996-2003 period to avoid years of high inflation, exchange rate and price adjustments during the turbulent period of economic transition in early nineties. The prices in the new member states before EU accession showed higher volatility compared to those in the established member states which reflects a relatively more liberal attitude towards farm policy compared to EU-15.

The scenarios with 'partial liberalisation' are assumed to enhance the magnitude of volatility in the EU but not to the level historically observed in the world market. In the liberal scenario it is assumed that full CAP liberalisation scenario involves simultaneous liberalisation of farm policies by other 'big players' in the world market which is likely to reduce price volatility in world markets overall. This would limit the increase in the volatility in the EU when opening its market. Other studies suggest that policy change may result in as much as a 21-45% reduction in the volatility of world market prices for cereals, the effect depending on variables such as the initial restrictiveness of policies towards price transmission relative to the size of the world market and on price elasticity (OECD, 2004b).

For modelling the indices of volatility, coefficient of variations were determined for each product (see Annex Figures B.5 and B.6) and member state, further used for adjusting initial volatility as measured in the 2002-2004 base period.

### *3.2.7 Cost assumptions*

Future inputs and costs were assumed on the basis of expert judgment. The assumptions reflect changes in input prices and take into account a variety of factors influencing each input market. For the most part costs are determined by forces beyond the agricultural sector, such as likely increases in energy prices, land lease costs, inflation of wage costs or interest rates. It was decided that there is no reason to vary costs assumptions with the policy scenarios. Farm types such as intensive livestock which have feed as a major cost are an exception since prices of feed depend largely on prices of raw materials produced within the agricultural sector, mainly grains and protein. Model cost parameters for future scenarios were extrapolated from base levels at cost specific rates, determined independently by experts from each member state.

### *3.2.8 Yields – future levels and volatility*

Future yield levels have been determined through extrapolation of past trends in the period 1992-2004 with some corrections based on member state experts' judgement on the pace of technological change, efficiency improvements and other factors in each sector and member state. A simplifying assumption on the neutrality of policy scenarios for yield levels and variability has been made. Thus an adverse effect of a decrease in farm support for yield improvement is deemed to be counterbalanced by induced improvements in efficiency and technology. Future volatility of yields in each policy scenario (as measured by the coefficient of variation) are assumed to be equal to that in the base period.

The analysis of historical data revealed low rates of yield improvements in Poland and Hungary which can be attributed to a variety of unfavourable (both financial and structural) conditions related to the economic transformation. The relatively low current yields and general improvement in economic conditions due to EU accession, suggest that growth rates above those calculated from historical trends should be applied in most cases due to a likely 'catching-up' process.

Similar pattern of changes in the rate of yield increases applies to Spain, although in the case of some crops experts indicate that the existing trend of changes cannot be continued with the same speed. For Germany and Netherlands, in general, future yield predictions are based on a simple extrapolation of past trends. In the case of the Netherlands for some crops the yield increase rates were reduced by experts, due to the fact that yield levels are already the highest among considered member states and close to the levels attainable with existing technologies. Examples of assumed rates of yield increases and forecast yield levels are presented in Table 3.4.

**Table 3.4.** *Assumed rates of annual yield increase and yield forecast for selected commodities.*

| | Yields | Wheat | Rye | Barley | Potatoes | Sugar beet | Oilseed rape | Milk |
|---|---|---|---|---|---|---|---|---|
| Germany | Rate of increase | 1.5% | 1.9% | 1.33% | 3.0% | 1.2% | 2.2% | 2.5% |
| | Mean 2002-2004 | 71.9 | 51.5 | 57.4 | 398.3 | 577.3 | 33.5 | 6,531 |
| | 2018 | 89.0 | 67.0 | 69.1 | 602.4 | 682.2 | 45.4 | 9,229 |
| Hungary | Rate of increase | 1.5% | 2.0% | 1.5% | 1.5% | 2.6% | 1.0% | 2.0% |
| | Mean 2002-2004 | 37.6 | 20.7 | 31.7 | 183.3 | 418.3 | 19.5 | 5,985 |
| | 2018 | 46.3 | 27.3 | 39.0 | 225.8 | 599.2 | 22.5 | 7,898 |
| Netherlands | Rate of increase | 0.5% | 1.0% | 1.0% | 1.0% | 1.0% | 0.5% | 0.97% |
| | Mean 2002-2004 | 84.8 | 49.3 | 59.9 | 438.7 | 611.7 | . | 7,560 |
| | 2018 | 90.9 | 56.7 | 68.9 | 504.2 | 703.1 | . | 8,650 |
| Poland | Rate of increase | 2.0% | 0.9% | 1.3% | 2.0% | 2.0% | 0.5% | 2.5% |
| | Mean 2002-2004 | 38.4 | 24.5 | 31.7 | 189.3 | 427.0 | 23.5 | 4,127 |
| | 2018 | 50.7 | 27.8 | 38.0 | 249.8 | 563.4 | 25.2 | 5,832 |
| Spain | Rate of increase | 1.5% | 2.0% | 2.0% | 1.8% | 1.8% | . | 3.0% |
| | Mean 2002-2004 | 29.9 | 17.3 | 29.5 | 276.3 | 669.7 | . | 5,355 |
| | 2018 | 36.8 | 22.9 | 39.0 | 354.7 | 589.7 | . | 8,089 |

The future volatility of yields has been set for all policy scenarios in the model at the base year level. It was assumed that historical variation of yields is mainly weather induced and independent of policy changes. The base year volatility (measured by standard deviation of coefficient of variation) have been calculated for each farm type from FADN time series.

## 3.3 Results

### *3.3.1 Income volatility*

Simulation results are represented as mean values of farm incomes (Table 3.5A-E) and the probability of low incomes (Table 3.6A-E). It is important to emphasise that the estimated future performance of modelled farms depends strongly on the assumptions made as well as the quality of the entry data, largely stemming from FADN.

Simulations results revealed that the model is very sensitive to price changes. The expected positive impact of structural changes in agricultural markets at the global level results in a projected positive development of most farm prices on the EU and world markets. Similarly, assumed productivity increases have a positive effect on financial performance of farms. This, in general, weakens the adverse implications of policy liberalisation.

The main hypothesis of an adverse effect of liberalisation, i.e. gradual decrease and in the most liberal scenario removal of market price support and direct support, on farm income (as well as on the probability of negative incomes) has been verified as true.

The pattern of changes in the level of farm incomes across scenarios is similar for all member states and farm types under study. In general, farm incomes increase in some member states (e.g. Poland) or farm types (TF501) in the Most Likely 2013 scenario, and even more in some cases in Likely A 2018, but further are visibly being reduced when liberalisation is assumed (Likely B, Liberal 2018).

The cuts in sugar beet and milk prices, changes in the form of direct support and modulation and also greater costs of production were largely compensated in the Most Likely and Likely A scenarios by compensation payments for milk and sugar, as well as assumed productivity increases.

The 2003 reform of the direct support system (decoupling) brings significant redistributive effects in the case of the regional model of SPS (single area payment

**Table 3.5A.** *Mean farm income (Euro) – Germany.*

| Farm types (TF14) | ESU class[1] | Scenarios | | | | | |
|---|---|---|---|---|---|---|---|
| | | Base | ML 13 | LikA 18 | LikB 18 | Lib 18 | Pro 18 |
| 13 | 1 | 2,508 | -3,468 | -3,267 | -4,687 | -8,638 | 1,144 |
| 13 | 2 | 4,175 | -4,397 | -3,537 | -6,885 | -15,104 | 2,915 |
| 13 | 3 | 23,406 | 5,397 | 7,064 | 2,881 | -23,541 | 25,129 |
| 13 | 4 | 67,861 | 4,037 | 8,408 | -20,765 | -110,412 | 74,869 |
| 14 | 1 | 7,025 | -2,644 | 1,047 | 477 | -3,854 | 7,538 |
| 14 | 2 | 3,143 | -4,234 | -2,104 | -4,188 | -9,481 | 3,590 |
| 14 | 3 | 23,169 | 8,694 | 13,840 | 11,029 | -2,585 | 29,190 |
| 14 | 4 | 62,642 | 37,984 | 61,630 | 52,716 | 12,799 | 94,116 |
| 41 | 2 | 21,071 | 17,959 | 22,949 | 15,657 | 7,083 | 35,115 |
| 41 | 3 | 55,957 | 56,225 | 68,446 | 54,721 | 40,728 | 90,328 |
| 41 | 4 | 77,061 | 83,331 | 111,632 | 78,187 | 35,935 | 164,562 |
| 50 | 2 | 36,573 | 44,686 | 48,046 | 42,802 | 42,852 | 50,315 |
| 50 | 3 | 81,810 | 96,010 | 104,660 | 102,277 | 94,726 | 112,511 |
| 50 | 4 | 212,550 | 240,218 | 280,792 | 265,890 | 238,220 | 284,962 |
| 81-82 | 1 | 4,579 | 4,134 | 5,353 | 3,316 | 828 | 5,225 |
| 81-82 | 2 | 21,099 | 11,814 | 16,166 | 12,197 | 3,971 | 21,915 |
| 81-82 | 3 | 82,130 | 80,291 | 84,207 | 79,784 | 72,799 | 104,452 |
| 81-82 | 4 | 98,043 | 53,234 | 81,883 | 48,989 | -25,759 | 146,720 |

[1] 1 – 8-16 ESU, 2 – 16-40 ESU, 3 – 40-100 ESU, 4 - >100 ESU.

in the region). Germany is the only member state in the sample where such a solution is being implemented. Consequently on cereal and crop and mixed farms (with historically higher rates of direct aid) farm incomes in the year 2013 fall noticeably. In Poland the results are strongly positively affected by the phasing-in of the EU direct payments, which are increased from the original 55%[5] of the negotiated rates up to 100% in 2013.

The Liberal scenario results in a sharp decline in farm incomes, however not as dramatically as one could have expected. This is due to a relatively small drop in prices, assumed decreases in feed costs (linked with a general price trend for cereals) and lower fixed costs due to the lower profitability of farming which also

[5] Including complementary national direct payments (top-ups).

**Table 3.5B.** *Mean farm income (Euro) – Hungary.*

| Farm types (TF14) | ESU class[1] | Scenarios | | | | | |
|---|---|---|---|---|---|---|---|
| | | Base | ML 13 | LikA 18 | LikB 18 | Lib 18 | Pro 18 |
| 13 | 1 | 19,147 | 29,603 | 29,543 | 28,211 | 19,882 | 33,990 |
| 13 | 2 | 30,669 | 46,573 | 47,654 | 45,432 | 26,600 | 58,714 |
| 13 | 3 | 62,891 | 66,543 | 65,181 | 63,658 | 17,428 | 86,036 |
| 13 | 4 | 267,471 | 344,839 | 296,973 | 138,960 | 149,235 | 322,341 |
| 41 | 1 | 9,796 | 9,933 | 8,694 | 7,985 | 1,677 | 11,409 |
| 41 | 2 | 24,367 | 20,421 | 17,461 | 14,933 | 1,951 | 28,978 |
| 41 | 3 | 9,645 | -30,770 | -46,733 | -53,405 | -65,828 | -9,026 |
| 41 | 4 | 137,799 | 7,136 | -60,762 | -239,643 | -270,596 | 156,355 |
| 50 | 1 | 7,752 | 4,384 | 3,601 | 3,155 | 2,285 | 4,580 |
| 50 | 2 | 12,269 | 1,443 | 1,869 | 1,878 | -877 | 1,156 |
| 50 | 3 | 15,604 | -4,048 | -15,089 | -18,966 | -19,467 | -11,721 |
| 50 | 4 | 148,417 | -207,778 | -278,171 | -330,115 | -252,800 | -264,193 |
| 81-82 | 1 | 14,815 | 19,330 | 20,267 | 19,473 | 12,939 | 21,706 |
| 81-82 | 2 | 28,665 | 30,824 | 33,268 | 29,969 | 13,086 | 39,803 |
| 81-82 | 3 | 48,781 | 70,371 | 73,996 | 70,267 | 33,866 | 87,511 |
| 81-82 | 4 | 250,285 | -517 | -86,265 | -411,987 | -384,232 | 94,551 |
| 60 | 1 | 6,720 | 9,889 | 9,757 | 9,458 | 4,378 | 10,092 |
| 60 | 2 | 3,198 | 2,911 | 1,002 | -148 | -9,566 | 3,971 |
| 60 | 3 | 18,033 | 10,409 | 2,958 | -3,202 | -25,373 | 11,602 |

[1] 1 – 8-16 ESU, 2 – 16-40 ESU, 3 – 40-100 ESU, 4 - >100 ESU.

causes a decrease of land lease costs, and some other fixed costs. This would partly compensate for reductions in farm receipts from sales and withdrawal of the direct support. The protectionist scenario, on average, results in farm incomes that approximate those from the 2004 base year period.

### *3.3.2 Risk of low incomes*

The risk of low incomes has been measured by the percentage of farms with a below zero level of farm income.

The obvious conclusion is, that in the more liberal scenarios as compared to more protective environment farms are more exposed to risk due to incomes decreases.

**Table 3.5C.** *Mean farm income (Euro)– Netherlands.*

| Farm types (TF14) | ESU class[1] | Scenarios | | | | | |
|---|---|---|---|---|---|---|---|
| | | Base | ML 13 | LikA 18 | LikB 18 | Lib 18 | Pro 18 |
| 14 | 2 | -11,006 | -18,632 | -18,856 | -20,214 | -22,903 | -22,429 |
| 14 | 3 | 3,351 | -11,793 | -10,853 | -12,632 | -15,591 | -13,130 |
| 14 | 4 | 50,920 | 17,167 | 33,459 | 20,136 | 11,514 | 32,368 |
| 41 | 2 | 848 | 4,632 | 6,673 | 2,247 | -379 | 7,287 |
| 41 | 3 | 31,564 | 40,522 | 44,143 | 31,903 | 19,612 | 49,994 |
| 41 | 4 | 74,333 | 94,353 | 105,768 | 78,495 | 47,655 | 116,863 |
| 50 | 2 | 41,893 | 66,555 | 77,001 | 69,700 | 67,744 | 77,299 |
| 50 | 3 | 87,526 | 143,201 | 154,653 | 134,721 | 121,471 | 156,514 |
| 50 | 4 | 216,043 | 364,599 | 391,359 | 334,811 | 310,627 | 386,751 |
| 81-82 | 2 | -846 | 2,178 | 4,726 | 345 | -2,470 | -280 |
| 81-82 | 3 | 3,419 | 8,434 | 12,368 | 5,273 | 1,724 | 8,057 |
| 81-82 | 4 | 11,004 | 9,268 | 19,940 | 1,506 | -17,696 | 16,757 |
| 44 | 2 | -21,212 | -21,715 | -22,088 | -23,024 | -23,214 | -24,602 |
| 44 | 3 | -58,581 | -58,977 | -60,235 | -63,419 | -66,573 | -66,084 |
| 44 | 4 | -24,772 | -1,836 | 4,812 | -12,804 | -16,045 | 12,600 |

[1] 1 – 8-16 ESU, 2 – 16-40 ESU, 3 – 40-100 ESU, 4 - >100 ESU.

The Likely A and even Likely B scenarios, which offer much lower level of market protection, as an assumed effect of the Doha round, and only modest reductions in direct payments do not represent, in general, very serious threats. There are, however, differences in a the range of low income probabilities depending on the member state and farm types.

In general, farms in Germany, the Netherlands and Hungary show a higher vulnerability to policy changes as compared to Poland and Spain. This can be explained by several factors, such as higher fixed costs, lower rates of productivity increases and, primarily, by much greater dependence of German, Dutch and Hungarian farms on direct payments in the initial, base year.

Crop farms, on average, are threatened by policy liberalisation more than livestock farms, as measured by the percentage of farms with a negative income as compared to the base and protectionist scenarios. This observation relates mainly to German and Dutch crop farms (farm types 13 and 14), which are characterised by the

**Table 3.5D.** *Mean farm income (Euro) – Poland.*

| Farm types (TF14) | ESU class[1] | Scenarios | | | | | |
|---|---|---|---|---|---|---|---|
| | | Base | ML 13 | LikA 18 | LikB 18 | Lib 18 | Pro 18 |
| 13 | 1 | 10,882 | 13,493 | 14,215 | 13,461 | 6,595 | 17,337 |
| 13 | 2 | 20,558 | 25,339 | 27,240 | 26,568 | 12,466 | 33,088 |
| 13 | 3 | 72,548 | 82,256 | 88,258 | 81,132 | 50,704 | 107,473 |
| 13 | 4 | 203,010 | 227,516 | 242,499 | 238,118 | 163,930 | 284,249 |
| 41 | 1 | 7,148 | 8,178 | 10,608 | 7,737 | 4,647 | 14,368 |
| 41 | 2 | 15,689 | 16,586 | 21,004 | 16,075 | 9,633 | 28,923 |
| 50 | 1 | 5,877 | 7,529 | 8,193 | 7,197 | 5,455 | 8,721 |
| 50 | 2 | 15,705 | 16,980 | 18,900 | 17,081 | 14,083 | 20,247 |
| 50 | 3 | 23,313 | 22,263 | 28,473 | 23,133 | 17,253 | 28,381 |
| 50 | 4 | 138,389 | 145,787 | 159,982 | 136,940 | 126,925 | 168,473 |
| 81-82 | 1 | 5,849 | 6,528 | 7,444 | 6,432 | 3,526 | 9,362 |
| 81-82 | 2 | 23,071 | 24,479 | 26,732 | 24,695 | 18,305 | 30,879 |
| 81-82 | 3 | 38,159 | 41,737 | 48,014 | 42,410 | 24,828 | 54,356 |
| 81-82 | 4 | 102,444 | 99,346 | 127,635 | 88,014 | 30,116 | 164,529 |
| 60 | 1 | 3,186 | 2,817 | 3,547 | 2,619 | 446 | 4,888 |
| 60 | 2 | 8,917 | 9,230 | 10,063 | 8,497 | 3,435 | 13,813 |
| 60 | 3 | 3,542 | 928 | 5,050 | -453 | -14,040 | 10,052 |
| 71 | 1 | 5,260 | 6,120 | 7,535 | 6,150 | 3,302 | 9,149 |
| 71 | 2 | 10,213 | 11,803 | 14,235 | 11,185 | 6,080 | 18,577 |
| 71 | 3 | 30,448 | 32,366 | 38,127 | 30,321 | 18,680 | 53,150 |

[1] 1 – 8-16 ESU, 2 – 16-40 ESU, 3 – 40-100 ESU, 4 - >100 ESU.

highest in the whole sample share of fixed costs in total costs of production within the range 60-70%. Value of fixed costs per hectare is also noticeably greater than in respective farm types in other member states. On the opposite, fixed costs on crop farms in Poland and Spain contribute significantly less to total costs, and farms area is bigger.

In general, in all member states, farms specialising in animal production (dairy – 41 and pork – 501) show considerably higher income stability across scenarios compared to crop (13, 14) and mixed (81-82) farms. This can be partly explained by the relative stability of prices and production, mainly of milk, and lower levels of direct aid. In particular dairy farms are marked by a low risk due to a relatively

**Table 3.5E.** *Mean farm income (Euro) – Spain.*

| Farm types (TF14) | ESU class[1] | Scenarios | | | | | |
|---|---|---|---|---|---|---|---|
| | | Base | ML 13 | LikA 18 | LikB 18 | Lib 18 | Pro 18 |
| 13 | 1 | 6,499 | 5,189 | 5,702 | 4,928 | -242 | 7,015 |
| 13 | 2 | 16,488 | 13,498 | 15,007 | 13,299 | 2,003 | 18,028 |
| 13 | 3 | 29,914 | 22,681 | 26,842 | 22,806 | -1,513 | 31,120 |
| 13 | 4 | 64,205 | 39,977 | 43,964 | 33,706 | -25,122 | 53,586 |
| 41 | 1 | -5,170 | -2,509 | 1,243 | -2,922 | -5,976 | 5,080 |
| 41 | 2 | 24,183 | 31,213 | 41,073 | 33,614 | 25,248 | 49,171 |
| 41 | 3 | 66,053 | 83,978 | 109,050 | 90,468 | 64,266 | 124,382 |
| 41 | 4 | 172,120 | 223,135 | 301,524 | 243,363 | 177,636 | 343,648 |
| 50 | 1 | 17,430 | 19,944 | 21,632 | 20,888 | 19,738 | 23,087 |
| 50 | 2 | 39,302 | 46,809 | 49,882 | 45,692 | 45,035 | 51,300 |
| 50 | 3 | 113,658 | 121,678 | 151,203 | 136,143 | 149,681 | 136,982 |
| 50 | 4 | 107,445 | 151,075 | 166,628 | 142,967 | 139,304 | 169,870 |
| 81-82 | 1 | 14,924 | 14,995 | 15,846 | 11,962 | 2,488 | 18,659 |
| 81-82 | 2 | 12,233 | 10,970 | 12,455 | 9,574 | -1,698 | 14,757 |
| 81-82 | 3 | 26,024 | 25,172 | 29,243 | 19,632 | -6,233 | 35,829 |
| 81-82 | 4 | 148,695 | 149,391 | 154,427 | 127,340 | 44,967 | 171,731 |

[1]1 – 8-16 ESU, 2 – 16-40 ESU, 3 – 40-100 ESU, 4 - >100 ESU.

lower volatility of revenues from milk production and a smaller share of marketable, more volatile crops.

Simulation results show a tendency, that farms of the smallest and largest economic size are more exposed to a risk of low incomes than mid-size (cluster 16-40, 40 – 100 ESU) farms. A possible explanation is that small farms generate low incomes even with favourable policy assumptions, and under the worst condition they easily fall into the category of farms with negative incomes. The opposite cluster of the largest farms (more than 100 ESU) may suffer under more liberal policies because of their dependence on inputs such as hired labour and a burden of high fixed costs.

**Table 3.6A.** *Risk of negative farm income – Germany.*

| Farm types (TF14) | ESU class[1] | Scenarios | | | | | |
|---|---|---|---|---|---|---|---|
| | | Base | ML 13 | LikA 18 | LikB 18 | Lib 18 | Pro 18 |
| 13 | 1 | 27% | 79% | 76% | 83% | 94% | 41% |
| 13 | 2 | 34% | 66% | 64% | 74% | 89% | 40% |
| 13 | 3 | 12% | 42% | 39% | 46% | 86% | 11% |
| 13 | 4 | 15% | 48% | 48% | 62% | 91% | 17% |
| 14 | 1 | 23% | 64% | 51% | 55% | 64% | 32% |
| 14 | 2 | 39% | 66% | 59% | 66% | 76% | 40% |
| 14 | 3 | 23% | 42% | 36% | 39% | 55% | 23% |
| 14 | 4 | 30% | 39% | 33% | 36% | 50% | 23% |
| 41 | 2 | 0% | 0% | 0% | 0% | 0% | 0% |
| 41 | 3 | 0% | 2% | 1% | 5% | 25% | 0% |
| 41 | 4 | 0% | 0% | 0% | 1% | 4% | 0% |
| 50 | 2 | 19% | 15% | 16% | 19% | 19% | 15% |
| 50 | 3 | 20% | 17% | 17% | 17% | 21% | 14% |
| 50 | 4 | 17% | 15% | 13% | 16% | 21% | 13% |
| 81-82 | 1 | 27% | 31% | 29% | 35% | 46% | 26% |
| 81-82 | 2 | 18% | 36% | 29% | 36% | 48% | 23% |
| 81-82 | 3 | 12% | 17% | 16% | 19% | 22% | 10% |
| 81-82 | 4 | 19% | 35% | 25% | 36% | 61% | 10% |

[1] 1 – 8-16 ESU, 2 – 16-40 ESU, 3 – 40-100 ESU, 4 - >100 ESU.

**Table 3.6B.** *Risk of negative farm income – Hungary.*

| Farm types (TF14) | ESU class[1] | Scenarios | | | | | |
|---|---|---|---|---|---|---|---|
| | | Base | ML 13 | LikA 18 | LikB 18 | Lib 18 | Pro 18 |
| 13 | 1 | 1% | 0% | 0% | 0% | 4% | 0% |
| 13 | 2 | 5% | 2% | 2% | 3% | 18% | 1% |
| 13 | 3 | 6% | 7% | 11% | 12% | 43% | 4% |
| 13 | 4 | 9% | 13% | 22% | 39% | 41% | 15% |
| 41 | 1 | 5% | 5% | 10% | 12% | 27% | 4% |
| 41 | 2 | 4% | 6% | 8% | 14% | 28% | 4% |
| 41 | 3 | 8% | 4% | 4% | 7% | 16% | 5% |
| 41 | 4 | 14% | 48% | 52% | 69% | 69% | 24% |
| 50 | 1 | 20% | 33% | 40% | 39% | 43% | 35% |
| 50 | 2 | 24% | 49% | 49% | 50% | 54% | 51% |
| 50 | 3 | 32% | 54% | 64% | 69% | 69% | 65% |
| 50 | 4 | 32% | 76% | 82% | 87% | 77% | 78% |
| 81-82 | 1 | 2% | 0% | 0% | 1% | 5% | 0% |
| 81-82 | 2 | 1% | 1% | 1% | 3% | 23% | 0% |
| 81-82 | 3 | 10% | 7% | 7% | 10% | 31% | 7% |
| 81-82 | 4 | 19% | 53% | 63% | 88% | 87% | 38% |
| 60 | 1 | 15% | 10% | 12% | 13% | 31% | 11% |
| 60 | 2 | 38% | 41% | 44% | 47% | 71% | 37% |
| 60 | 3 | 26% | 36% | 47% | 53% | 80% | 36% |

[1]1 – 8-16 ESU, 2 – 16-40 ESU, 3 – 40-100 ESU, 4 - >100 ESU.

**Table 3.6C.** *Risk of negative farm income – Netherlands.*

| Farm types (TF14) | ESU class[1] | Scenarios | | | | | |
|---|---|---|---|---|---|---|---|
| | | Base | ML 13 | LikA 18 | LikB 18 | Lib 18 | Pro 18 |
| 14 | 2 | 78% | 88% | 88% | 88% | 89% | 89% |
| 14 | 3 | 52% | 61% | 61% | 61% | 64% | 63% |
| 14 | 4 | 36% | 47% | 44% | 45% | 51% | 43% |
| 41 | 2 | 52% | 26% | 24% | 41% | 57% | 26% |
| 41 | 3 | 4% | 1% | 1% | 6% | 18% | 1% |
| 41 | 4 | 3% | 1% | 1% | 5% | 20% | 1% |
| 50 | 2 | 32% | 20% | 19% | 24% | 24% | 20% |
| 50 | 3 | 11% | 2% | 3% | 5% | 7% | 2% |
| 50 | 4 | 14% | 3% | 3% | 6% | 8% | 3% |
| 81-82 | 2 | 50% | 45% | 44% | 50% | 60% | 54% |
| 81-82 | 3 | 43% | 37% | 36% | 45% | 49% | 40% |
| 81-82 | 4 | 45% | 48% | 40% | 53% | 63% | 43% |
| 44 | 2 | 100% | 100% | 100% | 100% | 100% | 100% |
| 44 | 3 | 100% | 100% | 100% | 100% | 100% | 100% |
| 44 | 4 | 82% | 54% | 43% | 69% | 74% | 33% |

[1] 1 – 8-16 ESU, 2 – 16-40 ESU, 3 – 40-100 ESU, 4 - >100 ESU.

**Table 3.6D.** *Risk of negative farm income – Poland.*

| Farm types (TF14) | ESU class[1] | Scenarios | | | | | |
|---|---|---|---|---|---|---|---|
| | | Base | ML 13 | LikA 18 | LikB 18 | Lib 18 | Pro 18 |
| 13 | 1 | 3% | 1% | 1% | 2% | 15% | 1% |
| 13 | 2 | 7% | 5% | 4% | 5% | 20% | 3% |
| 13 | 3 | 2% | 1% | 1% | 2% | 10% | 1% |
| 13 | 4 | 1% | 1% | 1% | 1% | 3% | 0% |
| 41 | 1 | 1% | 0% | 0% | 2% | 14% | 0% |
| 41 | 2 | 0% | 0% | 0% | 2% | 13% | 0% |
| 50 | 1 | 14% | 11% | 10% | 12% | 21% | 8% |
| 50 | 2 | 8% | 8% | 8% | 13% | 16% | 6% |
| 50 | 3 | 21% | 22% | 20% | 26% | 29% | 20% |
| 50 | 4 | 9% | 10% | 9% | 14% | 17% | 7% |
| 81-82 | 1 | 2% | 2% | 2% | 3% | 16% | 1% |
| 81-82 | 2 | 0% | 0% | 0% | 0% | 1% | 0% |
| 81-82 | 3 | 2% | 2% | 2% | 3% | 15% | 1% |
| 81-82 | 4 | 6% | 8% | 5% | 14% | 37% | 3% |
| 60 | 1 | 10% | 16% | 12% | 19% | 45% | 7% |
| 60 | 2 | 4% | 4% | 4% | 6% | 26% | 2% |
| 60 | 3 | 41% | 49% | 40% | 51% | 77% | 30% |
| 71 | 1 | 1% | 1% | 0% | 1% | 13% | 0% |
| 71 | 2 | 1% | 1% | 1% | 3% | 14% | 0% |
| 71 | 3 | 1% | 1% | 0% | 2% | 7% | 0% |

[1]1 – 8-16 ESU, 2 – 16-40 ESU, 3 – 40-100 ESU, 4 - >100 ESU.

## 3.4 Conclusions and discussion

The formulation of future policy scenarios and the quantitative assessment of income risks depend on many debatable assumptions and thus limit the analysis and conclusions. While the theoretical links between policy, market and farm structure variables are well known, their quantitative estimation is much more difficult. This applies largely to the availability of estimates of future parameters, the complexity of the analysis, as well as the quality of the initial farm data. Apart from methodological limitations, an interpretation of the results is not an easy task considering, for instance, the number of factors influencing farm incomes including the various farming conditions in the member states selected for study.

**Table 3.6E.** *Risk of negative farm income – Spain.*

| Farm types (TF14) | ESU class[1] | Scenarios | | | | | |
|---|---|---|---|---|---|---|---|
| | | Base | ML 13 | LikA 18 | LikB 18 | Lib 18 | Pro 18 |
| 13 | 1 | 9% | 15% | 15% | 17% | 51% | 11% |
| 13 | 2 | 5% | 10% | 8% | 12% | 41% | 6% |
| 13 | 3 | 7% | 13% | 9% | 15% | 50% | 9% |
| 13 | 4 | 5% | 15% | 14% | 20% | 73% | 9% |
| 41 | 1 | 61% | 54% | 43% | 56% | 62% | 34% |
| 41 | 2 | 5% | 3% | 1% | 3% | 8% | 1% |
| 41 | 3 | 3% | 1% | 1% | 3% | 8% | 1% |
| 41 | 4 | 2% | 2% | 1% | 2% | 6% | 1% |
| 50 | 1 | 23% | 22% | 20% | 21% | 24% | 20% |
| 50 | 2 | 21% | 17% | 18% | 20% | 21% | 17% |
| 50 | 3 | 23% | 23% | 19% | 23% | 21% | 23% |
| 50 | 4 | 34% | 27% | 27% | 29% | 31% | 27% |
| 81-82 | 1 | 0% | 0% | 0% | 1% | 30% | 0% |
| 81-82 | 2 | 6% | 8% | 9% | 15% | 57% | 5% |
| 81-82 | 3 | 13% | 14% | 12% | 20% | 58% | 9% |
| 81-82 | 4 | 0% | 0% | 0% | 0% | 5% | 0% |

[1] 1 – 8-16 ESU, 2 – 16-40 ESU, 3 – 40-100 ESU, 4 - >100 ESU.

In general the results show the considerable dependence of incomes and income variability on the support of the CAP on EU farms. This dependence varies greatly however across farm types, size and the member states. This is because of differences in terms of production patterns, changes in support levels, farming sector structures, inputs used and differences in the volatility of different commodity prices.

The simulation results show, that at least part of the agricultural sector is relatively immune to further liberalisation of market and farm support policies (Likely A and Likely B scenarios for 2018). With today's positive prospects for the future world market situation there is no significant change in the income levels compared to the Base 2004 and Most Likely 2013 scenarios provided direct support is maintained. Furthermore, under Likely A 2018, a more protective scenario, farm incomes on the most of the farm types in all member states improve, largely due to predicted productivity increases. Reductions in direct support (greater modulation and ceiling) assumed in the Likely B 2108 scenario gives more visible decreases in incomes, but not too different from the base situation.

Even complete market liberalisation proves not to be lethal for EU farming, although selected farm types in some of the member states analysed would be affected stronger than the others.

An adverse effect of liberalisation is the greater risk of low incomes as measured by the percentage of farms with a negative farm income, which is a consequence of both the lowering of agricultural commodity prices due to diminishing price support and increased variability of yields and prices as assumed in the model. This finding applies mainly to the Likely B and Liberal scenarios under which agricultural subsidies are significantly reduced or completely (Liberal) removed. Even under the Protectionist 2018 scenario, which assumes a return to a high support, Agenda 2000 type of agricultural policy, the risk of low incomes is greater than in the base year (2004) which can be explained by increased variability but also unfavourable trends in prices and cost developments over a long-time period.

The increase in the risk of low incomes under more liberal scenarios varies between farm types, depending on production type and economic size. The level of risk on specialised dairy and pig farms is still less significant than on crop farms. This is very likely due to the relative stability in productivity of livestock and the restricted exposure of animal farms to mainly market related risks.

There are some indications that the risk of low incomes increases significantly on some of the smallest farm types (8-16 ESU) and also on large farms too (above 100 ESU). It seems that the inefficiencies of small farms in less favourable and more turbulent economic environments are exposed. In the case of the largest farms a hypothesis is, that among other factors, the high share of fixed costs and especially hired labour are expected to increase greatly by 2018 and so are responsible to a large extent for the increased risk. Initial conclusions regarding the risks of low incomes in relation to farm economic size should be treated cautiously because the available FADN data was incomplete for the cluster of the largest farms. In some member states and farm types these farms were not represented at all in the FADN sample (e.g. there were no data on dairy farms in Poland of the size 40-100 ESU and above 100 ESU). It is also likely, that on some farm types the number of large farms in the sample was relatively small and the average parameters calculated for modelling were unrepresentative.

A general conclusion can be that any common income stabilisation schemes for the EU as a whole will need to be sensitive to the vulnerability of the different types of farms in different member states. Income variability and the risks of loss are shown to increase(in some cases quite dramatically) with increasing levels of liberalisation but to markedly different degrees. The lesson seems to be that there are internal strategies for farms of all types to moderate any increased price risks including structural adjustments. The danger is that the introduction of income stabilisation measures might slow down the adoption of such practices. Perhaps somewhat surprisingly there are some farm types especially in Poland and Spain where even the most liberal scenarios do not produce much reduction in income levels or much greater risks of negative incomes.

The conclusions are valid for the assumptions made. Since the model is very sensitive to price changes, any noticeable drop of prices below the most recently projected levels would increase the risk significantly. It should also be emphasised that modelling results were estimated with the use of a static simulation model. Under less favourable conditions farmers would in reality be able to reduce, to some extent, the risks of low incomes through adequate adjustments.

# Chapter 4

# Evaluating EU risk management instruments: policy lessons and prospects for the future

*Alberto Garrido and Maria Bielza*

## 4.1 Objectives

The objective is to summarise the policy-relevant conclusions that emerge from a thorough review of available risk management instruments. It builds on relevant literature, on recent work commissioned by EU institutions (Parliament: Cafiero *et al.*, 2005, and the European Commission 2005, 2006a, 2006b) and on qualitative field work carried out in several EU member states. The distinction between business risks and catastrophic and crisis risks is briefly clarified. Member state data from the EC (2006a), combined with other sources and the authors' own findings, provide a snapshot of the current situation and of important trends. Subsequently the major findings of the literature about the advantages and disadvantages of the most common risk management instruments is elaborated on. Finaly, three proposals of the EC (EC, 2005), which have attracted most attention recently (Cafiero *et al.*, 2005; EC, 2006a), are discussed.

## 4.2 Business risks versus catastrophic and crisis risks

Risk is a facet of hazard and is endogenous to the ability to cope, not a fundamental concept by itself. Policies and risk management instruments target both aspects, in an attempt to reduce vulnerability to hazards. Defined in these terms, precise risk measurements are difficult to come by, because the errors in defining and measuring hazards and vulnerability multiply the errors associated with risk evaluations. For instance, Cafiero *et al.* (2005) contend that, after the 2003 CAP reform, EU farmers' income will be more stable, contrary to a widely held belief, which is grounded on the supposition that EU farm prices would be more volatile (Alizadeh and Nomikos, 2005; Antón and Giner, 2005).

Cafiero (2005) and the European Commission (2006a) use regional indicators of yield variability, droughts and other variables mapped in GIS. While these certainly convey an idea of the sources of the regional variability of yields on continental scale, they fail to draw a clear distinction between entrepreneurial and catastrophic risks. Very little about the actual vulnerability of farms can be learned from these maps, in part because farmers are generally speaking well prepared to cope with their usual risks. Few direct measurements and analyses of the income variability of European farms are available. Comparing the income variability of farms of various member states included in the FADN, it is shown in Chapter 2 that the largest and more competitive farms tend to experience larger revenue instability (both in absolute and relative terms) than smaller farms. More productivity and size may be associated with more exposure and income stability, but also with greater accessibility to credit and a wealth of risk management instruments.

Furthermore, no study is available that evaluates the short and long-term consequences of serious crises for EU farmers. While there is a notable lack of research on what policies are most effective in dealing with risks on EU scale, a great deal of research has addressed specific risks on a regional or even smaller scale. Unfortunately, typical business risks, which are less difficult to manage in the EU, have been paid much more attention than crises and disasters.

## 4.3 Risk management instruments in practice

### *4.3.1 Policy and risk management instruments across EU member states*

As the European Commission (2006a) clearly details, there is a great diversity of policy options, risk management instruments and initiatives among member states. During the past five years, a number of member states (notably Austria, France, Greece, Italy and Spain) have given a serious impulse to policies aimed at providing farmers with a safety net. Others, including UK, Germany and the Netherlands, still rely on *ad hoc* relief and catastrophe compensation, and have reinforced farmers' training programmes for coping with risks. This great diversity of measures results from at least three different factors: (1) the types of risks and hazards faced by member states' farmers differ widely, (2) the extent to which farm holdings have been consolidated and restructured, and (3) the various approaches of member states to help farmers to cope with risk.

Animal health policy is perhaps the only area in which the EU has developed a common approach to reduce risky diseases. Unlike other agricultural risks, contagious animal diseases have regional, market and even human health implications. One key strategy for the protection of European livestock and citizens is to intensify border controls and enforce traceability, animal identification and labelling (European Commission 2006b). While this recent evaluation indicates a number of strengths and positive views, there are areas which demand renewed efforts. In particular, the way in which *ad hoc* compensation schemes are co-funded by the EU and member states may create incentives for moral hazard both at farms and member states levels. As Mangen and Burrell (2003) show, the financial consequences for national farms on and off quarantine zones create winners and losers, depending on whether exports are banned and on the magnitude of the epidemic. Consumer welfare also depends on the severity of an outbreak and on its market implications.

Leaving aside the EU's initiatives in the area of animal health, the following variables have been proposed to represent member states' policy approaches[6]: (1) Percentage of *ad hoc* and fund payments over total agricultural output (including crop and livestock); (2) Percentage of insurance premium over agricultural output (including also crop and livestock premium).

As Figure 4.1 shows, member states' national policies stand between two extremes represented by the UK and Spain. Based on recent data, the UK has mostly relied on *ad hoc* payments while Spain relies on agricultural insurance. Note that, apart from these countries, most member states spent less than 1% on both insurance premium and funds/*ad hoc* contributions. The combined expenditure on *ad hoc* payments and insurance in the Czech Republic and Slovenia are among the highest in the EU. In the following sections, we sharpen the focus on the differences among member states in the area of *ad hoc* payments and insurance.

According to data from the EC (2006a), almost all member states make *ad hoc* payments and a smaller percentage have either public or private stabilisation funds. *Ad hoc* payments to livestock growers are common in UK, Ireland, Belgium, Poland, Germany and Sweden (not represented in Figure 4.1 because of lack of insurance data). *Ad hoc* payments to crop farms are mostly related to frost, drought, hail and excessive rainfall. In France and Germany droughts have taken more than 65% and 30% respectively of the *ad hoc* payments made in the last ten years. On the basis of the available data, total annual *ad hoc* payments in the EU are about €1 billion (considering a period that varies among member states).

The EU has played a leading role in promoting animal health during the past decade (Council Decision 90/424/EEC). It has financed losses caused by animal disease using *ad hoc* compensations by means of market support instruments, and loss-based

[6] Unless noted all data used in this section is borrowed from (European Commission 2006a), Eurostat and complemented by the authors' own fieldwork.

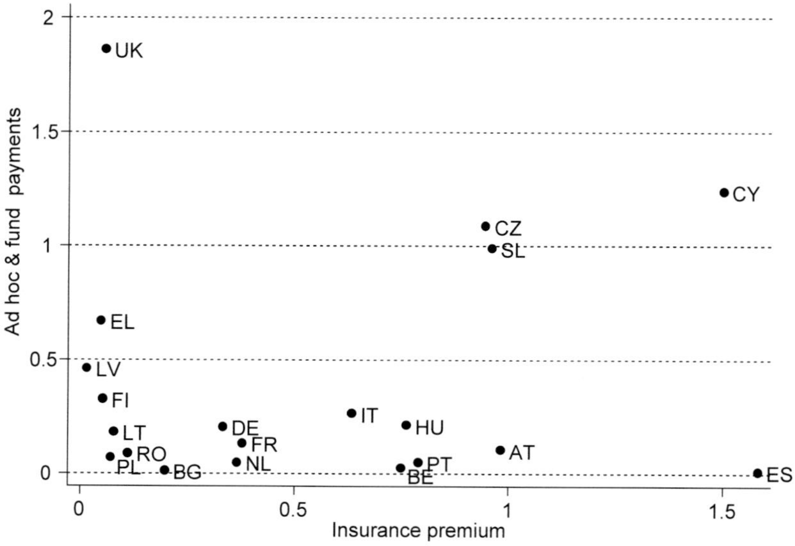


**Figure 4.1.** Ad hoc *payments versus insurance (annual payments expressed in % of total agricultural production). Source: European Commission (2006a), Eurostat and own field work. Length series vary across countries.*

compensation, using the 'veterinary fund'. The veterinary fund is fed by livestock farmers and/or member states' contributions (it varies across member states). The total budget for veterinary measures under Decision 90/424/EEC has peaked to €563 million in 2001 (with 80% assigned to the emergency fund for veterinary complaints, and 20% to disease eradication) and fell to €220 million in 2005 (with 91% and 2.2% in the same programmes) (European Commission, 2006b).

### *4.3.2 Agricultural insurance*

Agricultural insurance is offered in the EU in a wide variety of formats and with a wide degree of public-sector involvement. It is also evolving, with some member states such as Austria, Italy, France and Spain showing significant growth in the past few years. European Commission document EC (2006a) has offered the most detailed compilation since those authored by Forteza del Rey (2002) and OECD (2000b).

The major findings of these sources can be summarised by looking at the proportion of insured production and the impact of premium subsidies on some other key parameters. In Figure 4.2 we plot the proportion of insured crop production and the proportion of insured animal value against livestock production in most member

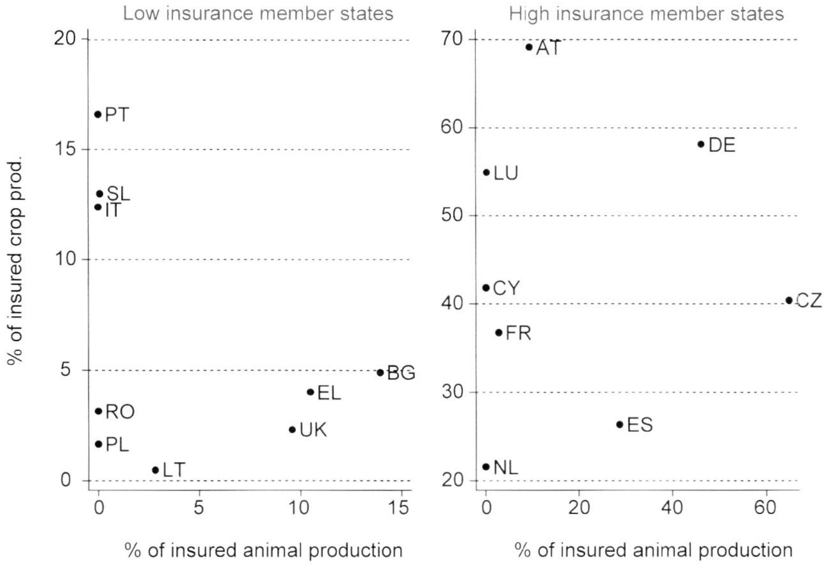


**Figure 4.2.** *Proportion of insured animal and crop value. Source: European Commission (2006a), Eurostat and own field work. Length series vary across countries.*

states for which data are available. In the left panel, we expand the scale to allow for a better representation of member states with low insured production. Note that percentages vary significantly across member states, with Austria, Germany, Denmark and the Czech Republic with the largest insured proportion of output. In the middle group, we find France, Cyprus and Spain.

Insurance policies vary significantly across member states, more than Figure 4.2 suggests. There are a number of factors that help us to understand these differences.

First, agricultural insurance exists both with and without the governmental support in the form of premium subsidies. Furthermore, even if a country has a high insurance penetration rate among crop and livestock farmers, that does not mean that all crop or animal risks are covered. In Figure 4.3, we plot the situation as regards to three key variables. The vertical axis represents premium subsidies measured as a percentage of total premium. The horizontal axis measures the ratio of premium to total insured production. The size of the circle represents total premium against total agricultural production.

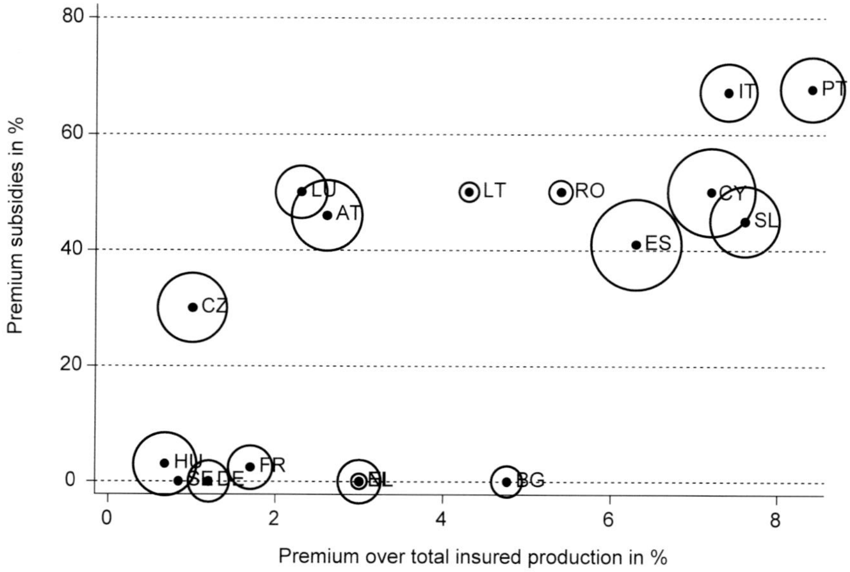


**Figure 4.3.** *Insurance policies across EU member states (Subsidies as a percentage of premium and premium as a percentage of insured production). Source: European Commission (2006a), Eurostat and own field work. Length series vary across countries.*

Figure 4.3 helps identify three groups of member states. There is the group of Mediterranean countries (except Greece, not shown because of the lack of information) in the upper right-hand side. These countries subsidise intensely the premium, and premium are relatively large with respect to total agricultural output. At the other extreme, we find Germany, Denmark, France, Ireland and Sweden. These countries' premium and subsidies are relatively small or zero. Total premium are also relatively small compared to the value of farm production. The size of the circle represents the percentage of total agricultural production that is normally insured. Data show that penetration rates are greater in countries where insurance is less subsidised, though coverages are broader in the Mediterranean countries.

Figures 4.1, 4.2 and 4.3 portray a landscape at EU level that can be summarised by the following trade-off. In order to provide a safety net, member states can resort to *ad hoc* payments in lieu of catastrophes or crises, however differently they may be defined around the EU, or else they can use subsidised insurance to promote the purchase of wide-coverage premium. Data from individual member states suggest that unless insurance is subsidised, coverages are limited and rarely cover crisis losses. When these occur, government expenditure in the form of *ad hoc* payments is unavoidable. Standing between these two policy extremes, member states such as France, Italy or Austria are swaying in the direction of more insurance. France and Spain have linked the eligibility to *ad hoc* funds to the purchase of agricultural insurance.

## 4.4 Major institutional trends and innovation

The previous brief review is no more than an updated snapshot of the situation around the EU. Field work carried out for this study shows that many countries are making progress and finding new ways of providing safety nets for their farmers.

### *4.4.1 Technical Innovations*

- Index insurance based on vegetation indices to cover drought episodes (USA, Spain, Canada and experimentally in France, South Africa and Ukraine).
- Parametric insurance (based on rainfall indices in Australia; water, rainfall and drought in India; rainfall indices in Morocco and Romania; this under study).
- Derivatives (river flow derivatives in Mexico for water supply risks).
- Contractual agriculture in speciality crops, using (1) forward contracting; (2) downstream-upstream contracting along the market chain; (3) derivatives and over-the-counter contracts (4) bankruptcy swaps in the rural banking sector (throughout the EU and the USA).

### *4.4.2 Institutional innovations*

- Austria, Italy and France are promoting and expanding the penetration of agricultural insurance.
- France began offering an experimental pilot revenue insurance policy for oil crops denominated in MATIF future prices.
- In Spain, Italy and France, the eligibility for receipt of *ad hoc* aid, as well as the size of such aid, is becoming increasingly conditional on purchasing agricultural insurance. *Ad hoc* aids are only granted to non-insurable hazards. The Netherlands cancelled the programme of compensation for excessive rainfall, which led to the launch of a public-private insurance.
- The UK is focusing on helping farmers diversify and on reinforcing training programmes in risk management. A proposal called '*Cost-sharing*' is trying to set up an arrangement to share the costs of fighting disease outbreaks between government and industry. This scheme would consist of a system of taxes which would accumulate from year to year, with some similarity to a fund.
- The Netherlands and UK do not make *ad hoc* payments to compensate for climatic hazards.

- The USA now offers on-line probability estimates of single and multiple climatic hazards for various time-scales (days, weeks, months) on a very small geographical scale. This offers option contracts and derivatives a chance to deal with very specific climate protection products (Turvey, 2007).

## 4.5 Results

### 4.5.1 About agricultural insurance

The history of agricultural insurance is full of bad examples and critical periods (see Hueth and Furtan, 1994; Hazell *et al.*, 1986). However, there are many other positive evaluations (Burgaz Moreno and Pérez-Morales, 1996; Mishra, 1996). Furthermore, many problems are related to asymmetric information, and a number of problems have been ascribed to public participation in agricultural insurance[7]:

### 4.5.2 Lack of competition in the insurance market

To date, no rigorous study has evaluated the degree of market power and lack of competitiveness. Yet, as premium are set by the insurance companies, competition among companies can only be based on their retailing services, and the additional insurance or financial services that they offer farmers. In most cases, agricultural insurance is just a small part of the insurance business contracted in rural areas. The lack of competitiveness strictly attributable to agricultural policies is likely to be small.

Using data collected by the EC (2006a), statistical analyses were performed in order to evaluate the extent to which competition in the insurance markets has some impact on various national insurance data. In Table 4.1, member states in which insurance rates are formed in competitive markets are separated from those in which rates are set by governments or where there is only one selling agency.

The comparison of means of the five variables reveals that the two groups differ significantly only in the percentage of insured crop production (line 2). The group with 'no competition' has a significantly higher penetration rate of insurance in crop production. In the remaining variables the group of member states with competitive insurance markets does not differ significantly from the other group.

[7] We are indebted to Professor Brian Wright for the ideas reviewed here.

**Table 4.1.** *The impact of competitiveness on key insurance parameters among EU member states (averages, standard deviation and means t-tests; n, number of countries included in each group).*

| | Without competition | With competition | Means comparisons (T-tests) |
|---|---|---|---|
| 1 % Of insured animal production | 7.45 (10.4)<br>n=7 | 7.65 (13.9)<br>n=18 | p>0.1 |
| 2 % Of insured crop production | 28.03 (28.06)<br>n=7 | 10.41 (12.3)<br>n=18 | p<0.05 |
| 3 % Premium over insured production | 4.3 (2.1)<br>n=6 | 3.7 (3.1)<br>n=11 | p>0.1 |
| 4 Loss ratios | 0.77 (0.22)<br>n=6 | 0.71 (0.27)<br>n=13 | p>0.1 |
| 5 Type of insurance (1,2,3) [1] | 2.14 (0.9)<br>n=7 | 1.72 (0.67)<br>n=18 | p>0.1 |

*p<0.05.

[1]Categories are 1 (only if single-risk insurance); 2 (if 1+ MCPI insurance) and 3 (if 2 + yield insurance).

Source: European Commission (2006a), Eurostat and own fieldwork and statistical analysis. The list of member states included in the analysis are (a) Without competition: AT, CY, EL, IE, LT, LU, ES; (b) with competition: BE, BG, CZ, DK, FI, FR, DE, HU, IT, LV, NL, PL, PT, RO, SK, SI, SE, UK).

### *4.5.3 Rent seeking by insurance companies*

It is often argued that insurance companies end up capturing the profits resulting from the premium subsidies (Hueth and Furtan, 1994). While there is no evidence to support this conclusion, the fact that subsidies explain the growth of insured crops in all countries is an indication that insurance companies are surely the recipients of part of the subsidies. While there is no case available of a private system providing broad insurance coverage to growers, there is evidence of more rapid innovation in publicly run systems than in privately run systems.

### *4.5.4 Excessive loading rates in comparison to banking services*

Hazell *et al.* (1986) compared the administrative costs of agricultural insurance in many countries, about 6% of the insured value, with those of other insurance sectors, 1.5%. In Spain, Agroseguro (SA), the pool of insurers, had a ratio of running costs over total liability of 0.3% in 1993, while in 2005 it was 0.25%. As new policies are based on indices, with a technological and IT platform and no need to perform on-site loss adjustment, administrative costs are bound to be reduced.

### *4.5.5 Large transaction costs needed to prevent problems associated with asymmetric information*

This criticism is supported in view of the results of the USA and Spain from the early 80s and many developing countries. More recently, actuarial imbalances have been brought closer in line with those needed to ensure financial sustainability. This has been possible because both of these countries have made an effort to screen the individual farmers' risks, collect more historical individual and zonal data, and expand the portfolio of coverages to increase risk pooling benefits (as Table 4.2 shows, loss ratios are only slightly higher in member states with premium subsidies than in member states without them).

### *4.5.6 Subsidised insurance crowds out other privately provided instruments*

Literature has not ascertained the extent to which subsidised agricultural insurance crowds out other risk management instruments offered by the private sector. A safe hypothesis would be that subsidised insurance retailed by private insurers may be offered in combination with other insurance products, such as life, buildings or machinery insurance. These products are heavily used by farmers in member states where crop insurance is not subsidised. What the literature does suggest is that CAP per hectare subsidies reduced the incentives of farmers to hedge with futures and options.

In those countries where insurance has been publicly developed and uptake rates are high, farmers are generally given the option of selecting from a wide menu of coverages and policy formats. Data on insurance participation show that, at least in Spain, very few farmers exhibit continuous and invariable insurance strategies (Garrido and Zilberman, 2007). This result suggests that farmers follow economic incentives, learn from their insurance experience and select their portfolio of instruments according to rational (or at least pseudo-rational) criteria.

Traditionally, insurance strategies have been combined with the use of financial instruments, to which insured farmers have enjoyed preferential access granted by rural banks. It is very likely that insurance makes some risks much more transparent and that insurance experience helps farmers dissociate the sources of risks to which they are exposed.

In an attempt to evaluate whether premium subsidies have an impact on some of the key insurance variables, we performed some statistical tests to differentiate those member states with insurance subsidies (AT, CY, CZ, FR, IT, HU, LV, LT, LU, PT, RO, SK, Sl and ES) from those without them (BE, BG, DK, EE, Fl, DE, IE, NL, PL, SE, UK), using the data compiled by the EC (2006a). In Table 4.2, we compare for those member states for which data are available the proportion of insured animal production, the proportion of insured crop production, the premium over insured production, loss ratios and insurance types.

When we control for whether or not member states subsidise premium, we find statistically significant differences in four of the five descriptive variables. The results show that premium subsidies help to increase the value of insured crop value (line 1) and are accompanied by larger relative premium (line 3). Furthermore, premium subsidies tend to be associated with greater insurance diversity and coverage (line 5). However, the percentage of insured animal production is not significantly different among groups of member states, and loss ratios are lower in countries without subsidies.

**Table 4.2.** *The impact of insurance subsidies (averages, standard deviation and means t-tests; n, number of countries included in each group).*

| | | With premium subsidies | Without premium subsidies | Means comparisons (t-tests) |
|---|---|---|---|---|
| 1 | % Of insured animal production | 7.8 (18.17)<br>n=11 | 7.29 (13.9)<br>n=14 | p>0.1 |
| 2 | % Of insured crop production | 22.5 (22.76)<br>n=11 | 6.3 (12.3)<br>n=14 | p<0.05 |
| 3 | % Premium over insured production | 4.6 (2.82)<br>n=5 | 2.32 (1.59)<br>n=11 | p<0.01 |
| 4 | Loss ratios | 0.79 (0.28)7<br>n=7 | 0.62 (0.16)<br>n=12 | p<0.05 |
| 5 | Type of insurance (1,2,3) [1] | 2.21 (0.18)<br>n=12 | 1.33 (0.14)<br>n=14 | p<0.1 |

[1]Categories are 1 (only if single-risk insurance); 2 (if 1+ MCPI insurance) and 3 (if 2 + yield insurance).
*p<0.05.
Source: European Commission (2006a), Eurostat and own fieldwork and statistical analysis. Countries included in the analysis are: (a) With subsidides AT, CY, CZ, FR, DE, HU, IT, LV, LT, LU, PL, PT, RO, SK, ES; (b) without subsidies BE, BG, DK, EE, FI, DE, IR, NL, SE, UK.

## 4.6 Conclusions

In the following tables we rate all the instruments that have been under discussion by the European Commission (2006a,b) and by Cafiero *et al.* (2005). The bases for our judgment are literature, documentation and private interviews. In Tables 4.3A and 4.3B we attempt to rate each family of instruments, based on a number of criteria. Ratings are merely illustrative of major trends and are based on the assumption that instruments are applied using the best actual practice. The policy options that are reviewed include the EC's (2005) three options; i.e. (1) insurance for natural disasters, (2) stabilisation funds and (3) provision of basic coverage against income crises; Cafiero *et al.* (2005) *ex-post* compensation for catastrophes and incentives for lower hazard reduction; the EC's (2006a) seven alternative insurance options, ranging from single-risk to revenue insurance and public reinsurance.

On the basis of the above ratings, it seems evident that no one instrument outperforms the others in all aspects. Furthermore, instruments that would be more appropriate for natural catastrophes will behave poorly for outbreaks of animal epidemics. The economic efficiency of public initiatives (welfare gains per € of expenditure) is greater with smaller coverages and guarantees. Low-coverage instruments also provide a screening device to set up eligibility conditions for *ad hoc* payments related to non-insurable risks. A growing number of member states are building on this type of conditional eligibility, encouraging the connection of different instruments, and offering dual approaches that include privately provided instruments and public safety nets. Instruments that can be administered at less cost, using IT technologies, indexing components, with no or little need for loss adjustment also have a better chance of offering more value at the lowest cost.

Finally, the instruments that seem to best complement privately offered instruments and that are more compatible with other EU policies also offer the most basic forms of protection. These include catastrophe insurance or single-peril insurance, policies which the private sector offers in many countries and which it could increasingly promote in the near future, especially if loss adjustment procedures can be made online or using IT.

The following ideas appear to have strong support from the literature and the experiences reviewed:

1. Contingent-state contracts, futures/options and other index derivatives are useful mechanisms. However, agriculture in the EU is extremely diverse under natural conditions, as in terms of risks and structural situations. Widely traded securities/assets that permit hedging risks will be difficult to develop, because

**Table 4.3A.** *Rating alternative risk management tools (1=minimum; 5=maximum).*

| | Discriminates between normal risks, crises and disasters[1] | Addresses risks of livestock epidemics | Acceptance by | |
|---|---|---|---|---|
| | | | Farmers | Insurers and other private agents |
| **EC (2005)-Option 1** (insurance for natural disasters only) | 5 | 1 | 2 | 1 |
| **EC (2005)-Option 2** (stabilisation funds) | 1 | 2 | 2 (varies across member states) | 1 |
| **EC (2005)-Option 3** (providing basic coverage against income crises) | 2 | 2 | 3 | 1 |
| **Cafiero (2005) alternative proposal** (for ad hoc crisis aids; only *ex-post* direct damage compensation) | 5 | 4 | 2 | 2 |
| **EC (2007) – EU-wide system of agricultural insurance:** | | | | |
| Single-risk or MPC | 2 | 1 | 3 | 3 |
| Yield insurance | 2 | 1 | 4 | 3 |
| Whole-farm yield Insurance | 1 | 3 | 1 | 2 |
| Income/Revenue insurance | 2 | 4 | 3 | 4 |
| Area index insurance (arable crops only) | 2 | 2 | 2 | 3 |
| Indirect-index insurance | 3 | 2 | 1 | 4 |
| Public reinsurance | 2 | 2 | 4 (to the extent that insurance becomes cheaper) | 5 |

[1] 1: poor discriminant; 5 strong discriminant.

Source: Authors' own elaboration based on Garrido and Bielza (2008).

**Table 4.3B.** *Rating alternative risk management tools (1=minimum; 5=maximum).*

| | Prone to welfare losses due to information asymmetries | Incentives for: Miss-reporting actual losses | excessive risk exposure |
|---|---|---|---|
| **EC (2005)-Option 1** (insurance for natural disasters only) | 1 | 1 | 3 |
| **EC (2005)-Option 2** (stabilisation funds) | 1 | 1 | 1 |
| **EC (2005)-Option 3** (providing basic coverage against income crises) | 3 | 2 | 4 |
| **Cafiero (2005) alternative proposal** (for crises ad hoc aids; only *ex-post* direct damage compensation) | 1 | 2 | 3 |
| **EC (2007) – EU-wide system of agricultural insurance:** | | | |
| Single-risk or MPC | 2 | 1 | 3 |
| Yield insurance | 2 | 2 | 2 |
| Whole-farm yield Insurance | 3 | 3 | 4 |
| Income/revenue Insurance | 2 | 3 | 3 |
| Area-index insurance (arable crops only) | 3 | 2 | 3 |
| Indirect-index insurance | 4 | 1 | 4 |
| Public reinsurance | 1 | 1 | 2 |

Source: Authors' own elaboration based on Garrido and Bielza (2008).

basis risks and trading costs will be a serious obstacle to take-up by farmers. Yet, as technological innovations enable the development of more diverse index instruments, a market may develop for these in the EU. At present, the use of financial instruments among farmers and even cooperatives is low.

2. Farmers would profit from a diversified set of risk-management instruments that should target multiple risk sources both within-farm boundaries and across the market chain from the farm-gate to the wholesale market. In highly capitalised agriculture, we are seeing major innovations in contractual agreements along

| Cost effectiveness (ΔU/public expend) | Compatibility with other EU policies | Complement (1) / substitute (5) with privately offered instruments | Vulnerability to rent seeking | Reliance on large reinsurance costs | Administrative complexity |
|---|---|---|---|---|---|
| 4 | 5 | 2 | 2 | 4 | 3 |
| 4 | 4 | 2 | 2 | 1 | 3 |
| 5 | 2 | 4 | 3 | 5 | 4 |
| 4 | 2 | 1 | 2 | 2 | 2 |
| 4 | 3 | 1 | 4 | 1 | 3 |
| 3 | 2 | 3 | 3 | 2 | 4 |
| 3 | 2 | 3 | 3 | 3 | |
| 4 | 2 | 4 | 4 | 3 | 5 |
| 2 | 1 | 4 | 2 | 4 | 4 |
| 2 | 2 | 2 | 2 | 2 | 4 |
| 2 | 2 | 1 | 2 | 3 | 3 |

the market chain that will enable professional farmers to externalise part of their risks.

3. When risk instruments are subsidised, it is a general rule that instruments with higher coverage and risk reduction potential come with lower subsidy efficiencies. More euros are needed in relative terms to increase risk reduction effects, once these are already large. Yet, in the case of insurance, reducing subsidies would likely be followed by lower rates of use of instruments.
4. OECD countries seem to have developed two alternative models to provide safety nets and risk management tools to their farmers. The keywords of

***Model 1*** are: training, competitiveness, liberalisation and compensation schemes for catastrophes and crises. For ***Model 2***: crop/livestock insurance, premium subsidies, gradual reduction of public compensation and increasing importance of insurance.

5. In the EU, Model 1 seems to be followed by Northern member states, whereas Model 2 is generally the approach of Mediterranean countries, although Austria's policy fits better with Model 2. These two models cannot easily converge to a middle ground mix. At most, member states are increasingly requiring that farmers contract insurance to become eligible to *ad hoc* compensation payments in case of crises or catastrophes.
6. Actuarial loss ratios of mature and growing agricultural insurance systems in the world have shown consistency and soundness. Actuarial techniques have improved significantly, helping countries control problems of asymmetric information and of poor loss adjustment procedures. Technologies, data mining, surveillance, and better risks evaluations explain these improvements. The era of poor insurance performance indicators around OECD countries came to an end in the mid-1990s.
7. Publicly provided (or publicly regulated and subsidised) crop insurance in OECD countries suffered from problems of asymmetric information in the 80s and early 90s. Now loss ratios of private insurance do not differ significantly from those of publicly provided insurance. On the EU scale, actuarial ratios do not differ significantly among member states with or without subsidised premiums.
8. OECD countries are constantly innovating and developing new instruments to underwrite or transfer risks and to provide new guarantees. Many of these are technology based and have great potential because administrative costs are much lower than in traditional crop insurance. Innovations are also of institutional nature, like new contractual definitions and design, market regulations and new modes of government participation.
9. Growing insurance portfolios increase the effects of risk-pooling and reduce the cost of reinsurance in relative terms. Some hazards, such as droughts or epidemics, for which disaster payments are offered in some countries, are now insurable (even though in most cases with the back of some sort of public reinsurance). In a near future, the trade of weather derivatives in the derivatives markets can further increase this effect and hopefully permit the private insurance of systemic risks.
10. While many working documents differentiate between normal risks and crises/catastrophes, past and existing policies cannot be equally categorised. There are countries whose definition of catastrophes encompasses hazards that are considered normal risks by others, and vice versa.

11. The area of contagious animal diseases is clearly the one that has been approached by the EU following a common policy. And yet, issues of moral hazard, both in the case of member states as well as farmers, subsist at the time of applying the protocols in cases of outbreaks. Further improvements are needed to ensure that the probability of crises in the livestock and meat sectors is lowered.

# Chapter 5

# Farmers' perceptions on risk and crisis risk management

*Peter Palinkas and Csaba Székely*

## 5.1 Objectives

As an important part of the 'Design and economic impact of risk management tools for European agriculture' research project, farmers' perceptions regarding (crisis) risk and (crisis) risk management have been surveyed in selected member states of the European Union. The process and results of this work are described in this document.

As a result of negotiations among the project participants five member states of the European Union were selected where agricultural producers were surveyed to elicit their perceptions regarding the issues under investigation. Because of practical considerations and the need for also including new EU member states in the research, the following member states were selected: Germany, Hungary, Poland, Spain and the Netherlands. Regarding the practical aspects of the research these member states were the most obvious choices as the partners participating in the project are located in these member states, so that the task of surveying the perception of agricultural producers was easier to organise and perform.

## 5.2 Materials and methods

To undertake the task, a questionnaire survey was selected as the applied research methodology. The questionnaire was designed to be completed in all of the selected member states, having been translated into the native languages of the given member states. The questionnaire was developed through a series of discussions among project participants, with due consideration given to relevant literature (Anderson *et al.*, 1977; Williams *et al.*, 1995; Hardaker *et al.*, 1997; Malhotra, 1999; EC, 2001; Agresti, 2002; Chambers and Skinner, 2003; Lehtonen and Pahkinen, 2004; EC,

2005; Kapronczai, 2005). The final version was accepted after six draft versions and a pilot survey to improve the quality of the questionnaire. The project participants agreed on a sample size of 200 farmers/producers to complete the questionnaire in each of the selected member states. In the end, all the member states supplied at least this quantity of completed questionnaires, in some cases even more (Hungary: 204; Poland: 206; the Netherlands: 236; Spain: 200; Germany: 201).

The selection of respondents followed a sampling plan and included a detailed description of selecting the farmers. Stratified sampling with proportional allocation was used as the sampling method for the questionnaire survey. Strata applied in the sampling plan for each member state were economic size of the holdings and their type of agricultural activity, both category groups were based on the FADN farm typology of the European Commission and the data source applied for the sampling was the Farm Structure Survey 2003, which was available for all selected member states at the time of constructing the sampling plan. Preparing the sampling plan helped us to establish representative samples for all the five member states under investigation (Lohr, 1999; EUROSTAT, 2003a,b, 2005; KSH, 2004; ATTRA, 2005).

Selected project partners from each member state included in the survey were responsible for organising the realisation of the survey in their respective member states. Final versions of the questionnaire were translated by the local project partners in each member state. In Hungary and Poland the survey was arranged through the national FADN institutions, in Spain through a survey company specialised in agriculture, in the Netherlands through an agricultural insurance company, while in Germany through a network of professionals having contact with relevant farmers. After the completion of the questionnaires the results were recorded in a computer file. After recording the data in the file, data was processed using a statistical software package. The time necessary for completing the questionnaires varied from member state to member state, but the predefined number of completed questionnaires was received from all selected member states.

This chapter presents the findings of the statistical analysis of data and is accompanied by an annex containing the tables relevant to the main text. The structure of the annex follows that of this chapter. Statistical analyses involved the exploration of overall differences among groups (primarily member states) followed by pair-wise comparisons of groups to elucidate the differences in more detail. In tables containing and comparing proportions of answers in member states, results are based on two-sided tests with significance level 0.05. For each significant pair, the key of the category with the smaller column proportion appears under the category with the larger column proportion. Tests are adjusted for all pair-wise comparisons within a row of each innermost sub-table using the Bonferroni correction. A similar approach applies to the comparison of interval data where overall differences among groups were revealed by the Kruskal-Wallis test at a significance level of 0.05, followed by a series of post-hoc tests, using the Mann-Whitney test, to explore differences in pair-wise comparisons of groups (using Bonferroni correction). Tables representing these results express the differences in a way that the key of the category with the significantly smaller mean appears under the category with the larger column mean. Data labelled 'Greater than' refers to these pair-wise comparisons in tables depicting interval data.

In relevant tables (Tables 5.2 and 5.3) the notation 'Valid cases' refers to the number of respondents who completed the given multiple response question correctly while 'n' in each column means the number of respondents within the valid cases who answered 'yes' for the given option within the set of possible responses.

Surveying farmers' perceptions on possible roles of financing partners was excluded from the research as the experience of pre-testing showed that it is hardly possible to receive reasonable feedback from farmers on this topic.

## 5.3 Results

### *5.3.1 Risk and crisis risk perception*

Farming activity is exposed to the influence of different important factors prevailing in agriculture. The effect of some factors may be either beneficial for farmers, for example political measures may provide better circumstances for them, or may also cause newly emerging problems. Farmers' subjective judgments on these factors determine also the resources and efforts devoted to offset the risks that may arise. In our survey farmers were asked to rate some of these influential factors according to their subjective opinions (Table 5.1). Factors could be rated from 1 (factor has no effect on farming) to 7 (factor has major effect on farming).

Overall averages show that weather and natural disasters are considered as the factors with the largest effect followed by volatility of prices. In the case of weather and natural disasters three groups of countries can be observed based on the statistically significant difference of the average rating of this factor. Polish and Hungarian farmers gave the highest ratings to this factor showing that weather and natural disasters have large effects on farming. Following them the Spanish average rating is somewhat lower but still referring to large effects. The third group consists of Germany and the Netherlands where according to farmers' opinions the

**Table 5.1.** *Rating sources of risk*
*(member state averages; 1-3=no effect; 3-5=moderate; 5-7=large effect).*

| | | Weather and natural disasters | Animal disease and epidemic | Price volatility |
|---|---|---|---|---|
| **Country** | | | | |
| Germany (A) | Mean | 5.41 | 3.35 | 5.35 |
| | N | 201 | 194 | 200 |
| | SD | 1.36 | 2.26 | 1.11 |
| | Greater than | - | - | - |
| Hungary (B) | Mean | 6.24 | 4.91 | 5.68 |
| | N | 204 | 89 | 202 |
| | SD | 1.09 | 1.86 | 1.27 |
| | Greater than | ACE | - | AC |
| Netherlands (C) | Mean | 5.06 | 5.98 | 5.24 |
| | N | 225 | 151 | 218 |
| | SD | 1.89 | 1.26 | 1.61 |
| | Greater than | - | ABD | - |
| Poland (D) | Mean | 6.41 | 5.19 | 5.55 |
| | N | 206 | 153 | 206 |
| | SD | 1.08 | 1.83 | 1.58 |
| | Greater than | ACE | - | - |
| Spain (E) | Mean | 5.74 | 3.36 | 5.48 |
| | N | 184 | 173 | 185 |
| | SD | 1.59 | 2.41 | 1.51 |
| | Greater than | AC | - | - |
| **Economic size of farm**[1] | | | | |
| Small (A) | Mean | 6.16 | 4.16 | 5.61 |
| | N | 325 | 214 | 324 |
| | SD | 1.32 | 2.36 | 1.38 |
| | Greater than | - | - | - |
| Medium (B) | Mean | 6.09 | 4.47 | 5.54 |
| | N | 212 | 159 | 210 |
| | SD | 1.30 | 2.16 | 1.56 |
| | Greater than | - | - | - |
| Large (C) | Mean | 6.25 | 5.02 | 5.49 |
| | N | 57 | 42 | 59 |
| | SD | 1.11 | 1.99 | 1.49 |
| | Greater than | - | - | - |

| Marketing difficulties | Input market | Debt | Political measures | Technological processes |
|---|---|---|---|---|
| 3.95 | 3.47 | 3.04 | 5.23 | 4.02 |
| 201 | 197 | 200 | 201 | 197 |
| 1.62 | 1.60 | 1.90 | 1.29 | 1.14 |
| - | D | B | BDE | D |
| 5.06 | 3.98 | 2.63 | 4.15 | 4.22 |
| 202 | 198 | 185 | 195 | 198 |
| 1.62 | 1.47 | 2.04 | 1.79 | 1.52 |
| ADE | ACD | - | D | D |
| 4.69 | 3.27 | 4.52 | 4.89 | 4.31 |
| 207 | 198 | 210 | 213 | 202 |
| 1.88 | 1.78 | 1.77 | 1.64 | 1.63 |
| AD | D | ABDE | BDE | DE |
| 4.05 | 2.21 | 3.42 | 3.31 | 3.64 |
| 206 | 204 | 205 | 206 | 200 |
| 2.04 | 1.50 | 1.87 | 1.63 | 1.79 |
| - | - | B | - | - |
| 4.39 | 3.75 | 2.97 | 4.07 | 3.62 |
| 181 | 179 | 179 | 183 | 183 |
| 1.94 | 2.19 | 2.15 | 2.15 | 2.21 |
| A | D | - | D | - |
| | | | | |
| 4.76 | 3.31 | 2.62 | 3.77 | 3.74 |
| 319 | 317 | 305 | 317 | 316 |
| 1.89 | 1.82 | 1.94 | 1.88 | 1.77 |
| BC | - | - | - | - |
| 4.25 | 3.28 | 3.41 | 3.73 | 3.91 |
| 211 | 207 | 206 | 209 | 207 |
| 1.94 | 2.03 | 2.05 | 1.92 | 1.98 |
| - | - | A | - | - |
| 4.00 | 3.18 | 3.72 | 4.52 | 4.05 |
| 59 | 57 | 58 | 58 | 58 |
| 1.81 | 1.97 | 2.06 | 1.84 | 2.00 |
| - | - | A | AB | - |

**Table 5.1.** *Continued.*

| | | Weather and natural disasters | Animal disease and epidemic | Price volatility |
|---|---|---|---|---|
| **Type of activity**[2] | | | | |
| Crop (A) | Mean | 6.06 | 2.24 | 5.54 |
| | N | 432 | 207 | 426 |
| | SD | 1.23 | 1.99 | 1.31 |
| | Greater than | B | - | B |
| Livestock (B) | Mean | 5.18 | 5.54 | 5.15 |
| | N | 226 | 227 | 224 |
| | SD | 1.83 | 1.66 | 1.63 |
| | Greater than | - | AC | - |
| Mixed (C) | Mean | 5.88 | 5.11 | 5.65 |
| | N | 300 | 300 | 299 |
| | SD | 1.37 | 1.77 | 1.29 |
| | Greater than | B | A | B |

[1] Includes Hungary, Poland and Spain where exact data was available. Small: 0-<8 ESU; Medium: 8-<40 ESU; Large: >40 ESU.

[2] Includes all selected member states. Crop: solely crop production; livestock: solely livestock production; mixed: both crop and livestock production.

effect of this factor is also considered large but at a lower level. These differences may be explained by the different or similar climatic features of countries.

Regarding volatility of prices, farmers of all selected member states share the same opinion, considering its effect as large but Hungarian farmers attribute larger effect to price volatility than Dutch and German respondents.

Animal disease and epidemics (where the farmer was involved in livestock production) is attributed as having large effects in Poland and the Netherlands, while the same applies to political measures in Germany, and to marketing difficulties in Hungary.

When considering the economic size of farms (based on ESU values) and type of farming activity similar conclusions can be drawn as in the other approaches, i.e. weather and natural disasters, and price volatility are identified as having the largest effect on farming. Not surprisingly, animal disease and epidemic have the

| Marketing difficulties | Input market | Debt | Political measures | Technological processes |
|---|---|---|---|---|
| 4.59 | 3.52 | 2.99 | 4.38 | 3.97 |
| 422 | 417 | 406 | 417 | 420 |
| 1.79 | 1.88 | 2.05 | 1.81 | 1.71 |
| B | B | - | - | - |
| 4.17 | 2.94 | 4.00 | 4.51 | 3.89 |
| 218 | 213 | 220 | 222 | 214 |
| 2.02 | 1.77 | 1.99 | 1.76 | 1.72 |
| - | - | AC | - | - |
| 4.43 | 3.36 | 3.41 | 4.29 | 4.05 |
| 300 | 295 | 294 | 299 | 288 |
| 1.81 | 1.75 | 2.00 | 1.86 | 1.59 |
| - | B | A | - | - |

largest effect according to livestock farmers followed by those who are involved in mixed production (including crop and livestock). As it was expected weather and natural disasters, and price volatility were perceived as having major effects on farming among those producing crops either solely or alongside livestock production.

### *5.3.2 Risk and crisis risk experience*

Although farmers try to reduce risks surrounding their activities, sometimes unexpected events may cause serious negative effects on their farming activities that may result in a crisis situation that often threatens farms with bankruptcy. Farmers were asked whether such a situation has ever occurred during their career. Spain proved to be the most critical member state as more than half of Spanish farmers (56.5%) had experienced a crisis situation so far, followed by Hungary with the rate of 40.3%. The other three member states can be considered as secure relative to Spain and Hungary (Figure 5.1 and Annex Table C.1A).

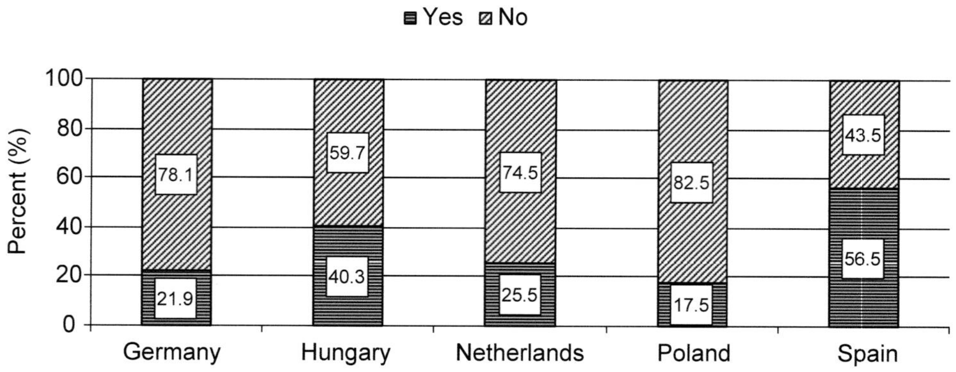


**Figure 5.1.** *Crisis experience among farmers. Source: Based on data taken from Annex Table C.1.A.*

Farmers were also asked to specify the main reasons of the experienced crisis situation. According to the answers, climatic (Hungary, Poland, Spain) and market conditions (the Netherlands and Germany) are the most important reasons in the selected member states (Annex Table C.1B).

To specify crisis frequency, yield/production loss was measured as the average number of times in the last ten years when unexpected loss exceeded 10% of planned yield and 5% of planned livestock production occurred (these thresholds are values agreed on by consortium members). Such yield losses occurred in Spain four times on average in the last ten years while in the other member states such events happened in three occasions (Annex Table C.2).

One approach applied to measure the magnitude of losses was to ask farmers to indicate the percent of cultivated land that was affected by the most critical yield loss. Considering member state averages it can be seen that in Spain, on average 69.58% of the land was affected by the specific loss, followed by Hungary with an average of 47.08% and Germany with 40% but the difference between these two member states was not statistically significant. The rate was 27.39% in the Netherlands and only 4.41% in Poland. It is also notable that in Spain the minimum proportion of the affected land was 33% while in the other member states the minimum value is zero and 2% in the case of Hungary. While in all other member states the maximum percentage of land affected was 100%, in Poland it was only 90%. The average percentage of total farm revenue affected by the most critical production loss was also measured. In terms of crop production the most critical yield loss affected on average 65.58% of the total farm revenue in Spain, followed by Hungary with 36.09% while in the other member states this value is under 30%, particularly in Germany which shows the lowest average value of 17.94% (Annex Table C.2).

To understand the background of yield losses that may result in crisis, factors harmful to crop production were rated by farmers based on the scale where 1 was harmless and 7 was very harmful. Overall averages show that viral and bacterial diseases are considered as the most harmful followed by fungi and insects. Regarding individual member states, viral and bacterial diseases are considered very harmful in Spain and Germany. In Hungary fungi, in Poland invertebrata and in the Netherlands insects received the highest ratings. Ratings indicate that these factors are very harmful (Annex Table C.3).

When analyzing these issues in livestock production, we found that losses exceeding 5% of planned production in the last ten years occurred only two times on average in all the five member states. On average, the percentage of livestock affected by the most critical production loss was the highest in Spain (49.6%) and then in Hungary (37.82%). In Germany and Poland the average magnitude of loss was between 16% and 18% but the difference between these is not statistically significant. The most critical production loss affected 44.52% of the total farm revenue in Spain, followed by Hungary (25.73%) and then the other three member states where this value is between 15% and 20% but the difference among these member states is not statistically significant. The maximum share of revenue affected is the highest in Spain with 100% while in the other member states it was considerably lower, between 50% and 70% (Annex Table C.4).

It can be concluded that on average the most critical losses appeared in Spain so far, both in crop and livestock production. Additionally, as an overall phenomenon it can be seen that the higher the share of land/production affected, the higher the proportion of total farm revenue affected (Annex Table C.5 and Table C.6).

### *5.3.3 Currently applied risk management instruments*

Besides knowing farmers' subjective perceptions on the effect of given factors and experiences related to risk or even crisis, the use of specific risk reducing methods applied by farmers is a highly important piece of information. Crop insurance is widespread in Germany and Spain where 60-70% of farmers apply this instrument which is more than in the other member states. The use of livestock insurance is significantly higher in the established member states (around 40%) than in the new ones. Marketing contracts are important in the new member states and Germany. German farmers are more active in off-farm investments (49.8%) and off-farm employment (36.8%) than those in the other member states. Property insurance is very important in Poland (67.5%), Germany (75.1%) and the Netherlands (66.8%). Avoiding the use of credit (maintaining a conservative debt ratio) is equally important in all member states (around 40%) while holding financial reserves is quite important in Hungary (40.5%), Poland (51.5%) and Germany (61.2%) unlike in the Netherlands (22.6%) and Spain (22.5%). Looking at individual member states, the situation is as follows. Hungary and Poland are identical in the sense that property insurance was indicated as the instrument applied by the highest percentage of farmers (41.5% and 67.5% respectively) followed by holding financial reserves (40.5% and 51.5% respectively). Avoiding use of credit is an important tool also in the Netherlands (38.1%). In Spain crop insurance was the primarily applied risk management instrument (59.2%) while this took a second place in Germany (68.7%) preceded by property insurance. In the case of Spain the second place was held by livestock insurance and avoiding the use of credit with 36.6% in case of both (Table 5.2).

If applied instruments are considered from the aspect of economic size and activity of farms the following findings are observable: larger farm size entails wider use of crop and livestock insurance; a higher proportion of smaller farms are employed also off-farm than larger farms as for the small ones farming alone may not provide a reasonable standard of living. Property insurance is the most widespread in both the farm size and activity approach while avoiding use of credit and holding financial reserves have quite similar positions, although it has to be noted that farmers running small farms are likely to try to run their business without credit as far as possible (Table 5.2).

### *5.3.4 Risk management instruments in the future*

Regarding farmers' future plans in terms of risk management instruments it can be observed that the majority of respondents in all member states, especially in Germany (80%) and Spain (75.1%), are willing to apply the risk management methods that they use now. In Poland farmers are more interested in other instruments than in the case of the other member states. In Spain, Hungary and the Netherlands farmers would like to avoid the use of credit while in Poland holding financial reserves was not attributed much importance. On the other hand, willingness of Polish farmers to try new instruments is unrivalled among the other member states. They are open to new, previously not used, solutions in managing risk. Many of them would like to be involved in crop insurance (41.3%), livestock insurance (37.4%), diversification (37.4%), off-farm employment (36.9%), vertical integration (33.5%) and hedging (58.3%) (Table 5.3).

When the economic size and the activity type approaches are considered the followings aspects are revealed. The majority of farmers, as in the member state comparison, would like to apply their already practiced instruments. Starting to use other new instruments is desirable only for smaller proportions of farmers in different size and activity groups. However, it is notable that with the increase of the economic size of farm, interest in getting involved in hedging also increases, up to around 30% of farmers of medium and large size holdings.

From another approach (tables not presented), among those farmers who would like to maintain the already applied methods, preferred new instruments would include holding financial reserves in Hungary (17%), the Netherlands (14.3%) and Germany (10%), hedging in Poland (60.3%), avoiding use of credit and accumulating financial reserves in Spain (18.2% in both cases). Among those farmers who would like to use some other methods instead of the currently applied ones (or desire to give up the already applied methods) the following instruments are the most prominent: crop insurance in Spain (55.1%), holding financial reserves in Hungary (47.5%), vertical integration (43.8%) and hedging (55%) in Poland, avoiding use of credit in the Netherlands (35.1%), and off-farm investment in Germany (47.5%).

### *5.3.5 Farmers' perceptions of risk reducing methods*

Based on their experiences and opinions farmers were asked to rate different risk reducing methods both in terms of crop and livestock production. Different methods could be rated from 1 (not effective) to 7 (very effective) and then member state averages were calculated. In crop production, based on overall averages it can be seen that preventive plant protection (5.5) was rated as the most effective instrument followed by technological improvement (5) and crop rotation/relay planting (4.7) rated third place. This means that preventive plant protection and technological improvement are perceived as very effective while the third method only rather moderately effective. Taking individual member states into consideration it can be observed that Hungarian, German and Dutch farmers consider preventive plant protection equally, as the most effective method of reducing risk of yield loss, and they also rated this instrument as very effective. The other group of member states consists of Spain and Poland where irrigation is selected as the most effective way of reducing yield risk and it was rated also as very effective based on the ratings given (Annex Table C.7).

In the case of livestock production, using the same scale, preventive medical treatment of animals (5.2) was rated the most effective based on overall averages. Ex-post medical treatment (5.1) was rated as the second, and young animals from own breeding (5) as the third most effective method of reducing risk of production

**Table 5.2.** *Current use of risk management instruments (Number and % of respondents using the instrument).*

| | | Crop insurance | Livestock insurance | Diversification | Marketing contracts |
|---|---|---|---|---|---|
| **Country** | | | | | |
| Germany (A) | N (valid=201) | 138 | 86 | 57 | 99 |
| | % of cases | 68.7% | 42.8% | 28.4% | 49.3% |
| | Greater than | BCD | BD | C | CDE |
| Hungary (B) | N (valid=195) | 42 | 8 | 45 | 75 |
| | % of cases | 21.5% | 4.1% | 23.1% | 38.5% |
| | Greater than | - | - | C | CE |
| Netherlands (C) | N (valid=226) | 69 | 84 | 26 | 42 |
| | % of cases | 30.5% | 37.2% | 11.5% | 18.6% |
| | Greater than | D | BD | - | - |
| Poland (D) | N (valid=206) | 29 | 14 | 69 | 73 |
| | % of cases | 14.1% | 6.8% | 33.5% | 35.4% |
| | Greater than | - | - | CE | CE |
| Spain (E) | N (valid=191) | 113 | 70 | 36 | 24 |
| | % of cases | 59.2% | 36.6% | 18.8% | 12.6% |
| | Greater than | BCD | BD | - | - |
| **Economic size of farm**[1] | | | | | |
| Small (A) | N | 78 | 33 | 70 | 79 |
| | % | 24.2% | 10.2% | 21.7% | 24.5% |
| | Greater than | - | - | - | - |
| Medium (B) | N | 80 | 42 | 58 | 74 |
| | % | 37.9% | 19.9% | 27.5% | 35.1% |
| | Greater than | A | A | - | A |
| Large (C) | N | 26 | 17 | 22 | 19 |
| | % | 44.1% | 28.8% | 37.3% | 32.2% |
| | Greater than | A | A | A | - |

| Production contracts | Off-farm investment | Off-farm employment | Property insurance | Vertical integration | Avoiding credit | Hedging | Holding financial reserves |
|---|---|---|---|---|---|---|---|
| 33 | 100 | 74 | 151 | 14 | 63 | 10 | 123 |
| 16.4% | 49.8% | 36.8% | 75.1% | 7.0% | 31.3% | 5.0% | 61.2% |
| E | BCDE | BCDE | BE | - | - | - | BCE |
| 31 | 8 | 37 | 81 | 7 | 74 | 3 | 79 |
| 15.9% | 4.1% | 19.0% | 41.5% | 3.6% | 37.9% | 1.5% | 40.5% |
| E | - | E | - | - | - | - | CE |
| 47 | 14 | 40 | 151 | 10 | 86 | 3 | 51 |
| 20.8% | 6.2% | 17.7% | 66.8% | 4.4% | 38.1% | 1.3% | 22.6% |
| E | - | E | BE | - | - | - | - |
| 33 | 4 | 42 | 139 | 12 | 83 | 6 | 106 |
| 16.0% | 1.9% | 20.4% | 67.5% | 5.8% | 40.3% | 2.9% | 51.5% |
| E | - | E | BE | - | - | - | CE |
| 11 | 11 | 9 | 57 | 24 | 70 | 2 | 43 |
| 5.8% | 5.8% | 4.7% | 29.8% | 12.6% | 36.6% | 1.0% | 22.5% |
| - | - | - | - | BC | - | - | - |
| 31 | 13 | 67 | 144 | 18 | 143 | 7 | 125 |
| 9.6% | 4.0% | 20.8% | 44.7% | 5.6% | 44.4% | 2.2% | 38.8% |
| - | - | B | - | - | BC | - | - |
| 30 | 9 | 16 | 104 | 16 | 71 | 0 | 78 |
| 14.2% | 4.3% | 7.6% | 49.3% | 7.6% | 33.6% | 0.0% | 37.0% |
| - | - | - | - | - | - | - | - |
| 14 | 1 | 5 | 29 | 9 | 13 | 4 | 25 |
| 23.7% | 1.7% | 8.5% | 49.2% | 15.3% | 22.0% | 6.8% | 42.4% |
| A | - | - | - | A | - | - | - |

**Table 5.2.** *Continued.*

| | | Crop insurance | Livestock insurance | Diversification | Marketing contracts |
|---|---|---|---|---|---|
| **Type of activity**[2] | | | | | |
| Crop (A) | N | 234 | 58 | 85 | 155 |
| | % | 53.9% | 13.4% | 19.6% | 35.7% |
| | Greater than | BC | - | - | B |
| Livestock (B) | N | 31 | 92 | 47 | 46 |
| | % | 13.7% | 40.5% | 20.7% | 20.3% |
| | Greater than | - | A | - | - |
| Mixed (C) | N | 91 | 102 | 91 | 95 |
| | % | 30.8% | 34.6% | 30.8% | 32.2% |
| | Greater than | B | A | AB | B |

[1] Includes Hungary, Poland and Spain where exact data was available. Small: 0-<8 ESU; medium: 8-<40 ESU; large: >40 ESU.

[2] Includes all selected member states. Crop: solely crop production; livestock: solely livestock production; mixed: both crop and livestock production.

**Table 5.3.** *Planned future use of risk management instruments (number and % of respondents using the instrument).*

| | | Same as now | Crop insurance | Livestock insurance | Diversification | Marketing contracts |
|---|---|---|---|---|---|---|
| **Country** | | | | | | |
| Germany (A) | N (valid=200) | 160 | 26 | 13 | 13 | 17 |
| | % of cases | 80.00% | 13.00% | 6.50% | 6.50% | 8.50% |
| | Greater than | CD | - | - | - | - |
| Hungary (B) | N (valid=202) | 141 | 30 | 5 | 18 | 34 |
| | % of cases | 69.80% | 14.90% | 2.50% | 8.90% | 16.80% |
| | Greater than | - | - | - | - | - |

| Production contracts | Off-farm investment | Off-farm employment | Property insurance | Vertical integration | Avoiding credit | Hedging | Holding financial reserves |
|---|---|---|---|---|---|---|---|
| 66 | 86 | 97 | 231 | 31 | 155 | 16 | 165 |
| 15.2% | 19.8% | 22.4% | 53.2% | 7.1% | 35.7% | 3.7% | 38.0% |
| - | BC | B | - | - | - | B | - |
| 28 | 11 | 32 | 146 | 12 | 82 | 1 | 77 |
| 12.3% | 4.8% | 14.1% | 64.3% | 5.3% | 36.1% | 0.4% | 33.9% |
| - | - | - | A | - | - | - | - |
| 53 | 37 | 68 | 173 | 21 | 117 | 5 | 146 |
| 18.0% | 12.5% | 23.1% | 58.6% | 7.1% | 39.7% | 1.7% | 49.5% |
| - | B | B | - | - | - | - | AB |

| Production contracts | Off-farm investment | Off-farm employment | Property insurance | Vertical integration | Avoiding credit | Hedging | Holding financial reserves |
|---|---|---|---|---|---|---|---|
| 12 | 28 | 21 | 15 | 9 | 17 | 8 | 32 |
| 6.00% | 14.00% | 10.50% | 7.50% | 4.50% | 8.50% | 4.00% | 16.00% |
| - | - | - | - | - | D | - | D |
| 20 | 18 | 21 | 14 | 8 | 32 | 1 | 53 |
| 9.90% | 8.90% | 10.40% | 6.90% | 4.00% | 15.80% | 0.50% | 26.20% |
| - | - | - | - | - | D | - | D |

**Table 5.3.** *Continued.*

| | | Same as now | Crop insurance | Livestock insurance | Diversification | Marketing contracts |
|---|---|---|---|---|---|---|
| **Country** (Continued) | | | | | | |
| Netherlands (C) | N (valid=214) | 140 | 21 | 29 | 16 | 25 |
| | % of cases | 65.40% | 9.80% | 13.60% | 7.50% | 11.70% |
| | Greater than | - | - | B | - | - |
| Poland (D) | N (valid=206) | 126 | 85 | 77 | 77 | 37 |
| | % of cases | 61.20% | 41.30% | 37.40% | 37.40% | 18.00% |
| | Greater than | - | ABCE | ABCE | ABCE | - |
| Spain (E) | N (valid=197) | 148 | 41 | 11 | 15 | 24 |
| | % of cases | 75.10% | 20.80% | 5.60% | 7.60% | 12.20% |
| Greater than | | D | C | - | - | - |
| **Economic size of farm**[1] | | | | | | |
| Small (A) | N | 213 | 76 | 38 | 50 | 50 |
| | % | 64.00% | 22.80% | 11.40% | 15.00% | 15.00% |
| | Greater than | - | - | - | - | - |
| Medium (B) | N | 156 | 67 | 46 | 48 | 37 |
| | % | 73.20% | 31.50% | 21.60% | 22.50% | 17.40% |
| | Greater than | - | - | A | - | - |
| Large (C) | N | 46 | 13 | 9 | 12 | 8 |
| | % | 78.00% | 22.00% | 15.30% | 20.30% | 13.60% |
| | Greater than | - | - | - | - | - |
| **Type of activity**[2] | | | | | | |
| Crop (A) | N | 327 | 84 | 26 | 58 | 61 |
| | % | 74.80% | 19.20% | 5.90% | 13.30% | 14.00% |
| | Greater than | B | - | - | - | - |
| Livestock (B) | N | 126 | 38 | 44 | 28 | 21 |
| | % | 58.10% | 17.50% | 20.30% | 12.90% | 9.70% |
| | Greater than | - | - | A | - | - |
| Mixed (C) | N | 216 | 68 | 61 | 49 | 38 |
| | % | 72.70% | 22.90% | 20.50% | 16.50% | 12.80% |
| | Greater than | B | - | A | - | - |

[1] Includes Hungary, Poland and Spain where exact data was available. Small: 0-<8 ESU; Medium: 8-<40 ESU; Large: >40 ESU.

[2] Includes all selected member states. Crop: solely crop production; Livestock: solely livestock production; Mixed: both crop and livestock production.

| Production contracts | Off-farm investment | Off-farm employment | Property insurance | Vertical integration | Avoiding credit | Hedging | Holding financial reserves |
|---|---|---|---|---|---|---|---|
| 22 | 23 | 19 | 36 | 7 | 42 | 13 | 42 |
| 10.30% | 10.70% | 8.90% | 16.80% | 3.30% | 19.60% | 6.10% | 19.60% |
| - | - | - | ABD | - | AD | B | D |
| 15 | 49 | 76 | 11 | 69 | 4 | 120 | 8 |
| 7.30% | 23.80% | 36.90% | 5.30% | 33.50% | 1.90% | 58.30% | 3.90% |
| - | BC | ABCE | - | ABCE | - | ABCE | - |
| 9 | 26 | 23 | 16 | 26 | 6 | 8 | 34 |
| 4.60% | 13.20% | 11.70% | 8.10% | 13.20% | 18.30% | 4.10% | 17.30% |
| - | - | - | - | ABC | AD | - | D |
| 31 | 66 | 63 | 23 | 59 | 46 | 52 | 56 |
| 9.30% | 19.80% | 18.90% | 6.90% | 17.70% | 13.80% | 15.60% | 16.80% |
| - | BC | - | - | - | - | - | - |
| 10 | 24 | 44 | 12 | 36 | 21 | 60 | 31 |
| 4.70% | 11.30% | 20.70% | 5.60% | 16.90% | 9.90% | 28.20% | 14.60% |
| - | - | - | - | - | - | A | - |
| 3 | 3 | 13 | 6 | 8 | 5 | 17 | 8 |
| 5.10% | 5.10% | 22.00% | 10.20% | 13.60% | 8.50% | 28.80% | 13.60% |
| - | - | - | - | - | - | A | - |
| 38 | 60 | 58 | 38 | 48 | 70 | 46 | 78 |
| 8.70% | 13.70% | 13.30% | 8.70% | 11.00% | 16.00% | 10.50% | 17.80% |
| - | - | - | - | - | C | - | - |
| 15 | 26 | 32 | 30 | 25 | 28 | 33 | 39 |
| 6.90% | 12.00% | 14.70% | 13.80% | 11.50% | 12.90% | 15.20% | 18.00% |
| - | - | - | C | - | - | - | - |
| 21 | 54 | 64 | 17 | 43 | 24 | 65 | 45 |
| 7.10% | 18.20% | 21.50% | 5.70% | 14.50% | 8.10% | 21.90% | 15.20% |
| - | - | A | - | - | - | A | - |

loss. On the other hand, it can be easily seen that these ratings are very close to each other so it can be assumed that they are considered equally effective. On the individual member state level, preventive medical treatment is termed as the most effective risk reducing method in Hungary, Poland and Spain and it was rated as very effective in all these member states. Ex-post medical treatment was found to be the most effective according to farmers in the Netherlands while in Germany quality assurance was rated as the most effective (Annex Table C.8).

### *5.3.6 Marketing channels applied by farmers*

Selling agricultural products through contracts or cooperatives is less risky due to provisional factors and conditions. Selling the products individually is probably the most risky way of marketing the products especially in the case of increased competition if the farmer lacks considerable bargaining power. The majority of Dutch (64.9%) and Spanish (53%) farmers stated they sell their products through cooperatives although individual sales in Spain (43.5%) are also important, as is also the case in Germany where 41.7% of the respondents stated that their primary marketing channel is individual sales. On the other hand, 40.7% of German respondents sell the majority of their products through cooperatives (Figure 5.2). In Hungary and Poland which are both new member states of the EU individual sales is still the most important marketing channel with 70% and 60% of respondents applying it respectively. There was no statistically significant difference detected between Poland and Hungary in this sense. The same applies to the group of the

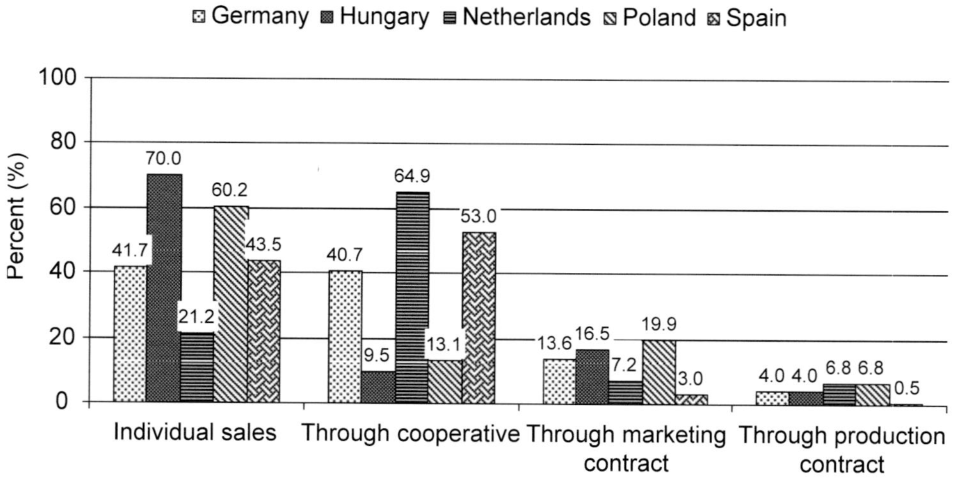


**Figure 5.2.** *Marketing channels applied by farmers Source: Based on data taken from Annex Table C.9.*

Netherlands, Spain and Germany regarding selling through cooperatives except for the Netherlands-Germany comparison where in the Netherlands a significantly higher proportion of farmers sell their products through cooperatives than in Germany (Annex Table C.9).

### *5.3.7 Financial aspects of farming*

Basically, farmers may obtain their revenues from two sources, from farming and from off-farm activities, which refers to revenue generating activities performed in addition to farming. In the case of the investigated member states German (60.7%) and Spanish (53.5%) farmers reported the highest rates of involvement in such activities, at a significantly larger proportion than in the other member states, while the difference between these two member states is not statistically significant. In the other three member states there is no statistically significant difference between the proportions of farmers having off-farm revenue. Between 30% and 41% of farmers in these member states indicated they obtain revenue from sources outside farming (Annex Table C.10).

To see how dependent farmers are on agricultural revenues, the average magnitude of revenue obtained from off-farm activities in relation to total revenue was measured. Based on farmers' responses, figures were close to each other in Hungary (38.26%), Poland (33.32%), Germany (34.19%), and the Netherlands (31.89%), while off-farm revenues were the highest in Spain where on average, 61.22% of the total revenue is gained from off-farm activities where this occurs. The average percentage of total revenue coming from off-farm activities in Spain is significantly higher than in the other member states while there were no statistically sound differences found between the member states other than Spain in this respect (Annex Table C.11).

Another, also very important aspect of off-farm revenue is whether it is continuously available for the farmer and thus represents a relatively stable source of finance or where it is only seasonal. In the latter case off-farm revenue may also be a great help for the farmer but it cannot be considered as the primary source of revenue. Regarding the distribution of off-farm revenue over time it can be observed that in all the selected member states at least 75% of farmers, who have such revenue, admitted that they earn off-farm revenue during the whole year. In country-country comparisons it turned out that there is no evidence of a statistically significant difference between the selected member states in this respect except for Spain and Hungary, where more of the relevant Spanish farmers have off-farm revenue throughout the year (Annex Table C.12).

The existence of debt refers to legal and financial obligations that may limit the decision authority of the farmer and also imposes extra risks because the debt has to be repaid within a certain period and thus deprives financial resources from farming activity. Using debt towards bank(s) to finance operations is widespread in Poland and the Netherlands, where 54%-65% of farmers have bank debt with no statistically significant difference between the two member states, although Dutch project partners noted that according to their opinion the Dutch data (54.2%) is not valid because they perceive in reality it is around 90%. The reason for this discrepancy was not revealed in this study. Germany, Poland and Hungary represent one group of countries as there was no statistically significant difference found between them. The share of farmers having bank debt currently in these countries is between 18% and 30% (Annex Table C.13).

Farmers were also asked to express their perceptions on the adequacy of access to credit. In the case of Spain and Hungary more than half of Spanish (50.3%) and Hungarian farmers (54.5%) stated that there is timely access to credit but only with strict conditions and high costs. There was no statistically significant difference found between these two member states. In Poland the largest group of farmers (41.3%) thought that costs and conditions of credit access are reasonable but requires a long procedure. In the Netherlands (81.3%) and Germany (78%) the majority of farmers stated that access to credit is timely and involves reasonable costs and conditions (no statistically significant difference detected). This may refer to the highly developed financial markets but it is surprising that this opinion of German farmers is accompanied by only a relatively low proportion of them having bank debt, although banks are not the only source of loans. It is worthwhile to mention that in Hungary 27.3% of farmers indicated that they have no access to credit at all, a rate which is considerably higher than in the other member states where this share of responses was only 1%-3% (Annex Table C.14).

### *5.3.8 Managing risk through assuring quality*

Besides governmental regulations, other standardised frameworks of rules (quality assurance methods/systems) can also be applied by farmers, mostly voluntarily. These applications help farmers to reduce production and market risk related primarily to the quality of products. In this respect Germany and the Netherlands are the overall leaders with around 80%-90% of farmers applying any type of quality assurance systems. The difference between these two member states is not statistically significant. The same applies to Spain and Poland where 68%-75% of farmers have such a system. Hungary is lagging behind with only 20.2% of farmers applying quality assurance system(s), which indicates an underdevelopment of the member state in this respect (Annex Table C.15).

### *5.3.9 Human resources risk*

With respect to up-to-date knowledge, attending professional educational courses is a very important way of obtaining valuable and directly applicable information. In Germany the vast majority, 76% of farmers have visited such a course recently while this rate is 61.8% in the Netherlands. The situation is different in the other three member states where only less than 30% of farmers attended such courses in order to keep themselves informed on the developments in agriculture – no statistically sound difference detected (Annex Table C.16).

In case of unexpected decline in farmers' health additional health insurance besides the obligatory one may provide extra compensation for possible financial losses and difficulties. Such insurance schemes are widespread in the Netherlands and Germany where 70.5% and 52.6% of farmers have any additional insurance of this type, respectively (Annex Table C.17). In the other member states such instruments are applied only by the minority, less then 25%, of farmers, without statistically significant difference among the individual member states (Hungary, Poland and Spain).

Buying life insurance schemes for the event of unexpected death is a somewhat more common practice also in all the examined member states, especially in Germany where 92.8% of farmers have life insurance. Germany is followed by the Netherlands with 62.2% of farmers having life insurance. In the remaining three member states this proportion is 30%-40% but the difference among these three member states is not statistically significant (Annex Table C.18).

### *5.3.10 Perception of institutional risk*

Besides measuring farmers' perceptions on diverse sources of risk and risk management methods we have also tried to reveal their opinions regarding the institutional environment surrounding them. We have distinguished two main dimensions of the institutional environment. On the one hand, farmers were asked to rate the extent to which the EU and national authorities, regulations (including financial subsidies) support their living, and on the other hand, how easy they find the adaptation to changes in the EU and national regulations. We have associated adverse opinion of farmers on these issues with higher risk perception, assuming that unfavourable influence of the institutional environment may result in greater difficulties for farmers. We found it also useful to consider what proportion of the respondents does not have opinions at all regarding the above-mentioned issues. Table 5.4 shows the main results of the statistical tests carried out on relevant data.

**Table 5.4.** *Perception of EU and national institutional environment.*

| Country | | Supportiveness of EU institutional environment[1] | Supportiveness of national institutional environment[1] | Adapting to changes in EU regulations[2] | Adapting to changes in national regulations[2] |
|---|---|---|---|---|---|
| Germany (A) | Mean[1,2] | 3.2 | 3.3 | 3.5 | 3.5 |
| | N | 177 | 174 | 177 | 174 |
| | SD | 1.2 | 1.2 | 1.3 | 1.3 |
| | Greater than | - | C | - | - |
| Hungary (B) | Mean[1,2] | 4.1 | 3.5 | 3.7 | 3.6 |
| | N | 187 | 196 | 191 | 197 |
| | SD | 1.1 | 1.1 | 1.4 | 1.3 |
| | Greater than | ACE | C | DE | D |
| Netherlands (C) | Mean[1,2] | 3 | 2.7 | 3.9 | 3.7 |
| | N | 194 | 203 | 202 | 207 |
| | SD | 1.4 | 1.3 | 1.3 | 1.4 |
| | Greater than | - | - | ADE | - |
| Poland (D) | Mean[1,2] | 3.9 | 3.5 | 3.4 | 3.4 |
| | N | 196 | 201 | 192 | 192 |
| | SD | 1.2 | 1.4 | 1.5 | 1.6 |
| | Greater than | ACE | C | - | - |
| Spain (E) | Mean[1,2] | 3.5 | 3.2 | 3.2 | 3.4 |
| | N | 200 | 200 | 198 | 198 |
| | SD | 2 | 1.6 | 1.7 | 1.5 |
| | Greater than | - | C | - | - |

[1] 1-3=not supportive; 3-5=moderately supportive; 5-7=very supportive.
[2] 1-3=very hard to adapt; 3-5=moderately easy to adapt; 5-7=very easy to adapt.

We have found that based on the statistically significant differences two groups of member states can be separated in the case of the judgment of farmers regarding the support they perceive from the EU. One group of countries consists of the new member states, Hungary and Poland, as farmers from these countries perceive the supportiveness of the EU in the same way as they consider it moderately supportive at the same extent. Farmers of the other country group, the established member states have a shared perception as they regard the EU institutions also moderately supportive, but at a lower extent than those from the new member states. In terms of expressing their opinions the Spanish farmers proved to be the most resolute as all of them provided an answer for the question. The case of Hungary and Poland is somewhat different but still less then 10% of farmers stated that they had no opinions on the issue. In Germany and the Netherlands approximately 12% of the farmers expressed no opinion regarding this issue.

In terms of the supportiveness of the national institutional environment a different grouping of member states can be observed. In Hungary, Poland, Spain, and Germany farmers consider the relevant national institutions as moderately supportive while Dutch farmers had a less favourable opinion in general, as they perceive such institutions as rather not supportive. However, it is also notable that no Hungarian and German farmers rated national institutions supportive at the highest possible level (rating it as 7 on the scale). In terms of no opinion Spanish farmers were as categorical as in the previous case and all of them expressed their opinions. In case of the other member states just a small minority of farmers from the new member states had no opinion (between 2% and 3%). Similarly to the previous question a considerably higher proportion of farmers expressed no opinion in the Netherlands and Germany.

Regarding how easy farmers perceive the adaptation to changes in EU regulations, generally Hungarian and Dutch farmers have a more positive perception than farmers from other member states and consider the adaptation process moderately easily. In case of the other three member states this process is perceived by farmers also as moderately easily but at a slightly lower level. In terms of no opinion we can observe that between 6% and 11% of farmers in the studied member states (with the exception of Spain) expressed that they had no opinion on this issue.

Additionally, farmers were asked to evaluate the ease by which they were able to adapt to changes in national regulations as well. Results show that generally farmers in all selected member states consider the adaptation process as moderately easy in terms of national institutions, and there is no statistically significant difference among the selected member states regarding this question. The only statistically significant but still small difference is between Hungary and Poland where the average Hungarian rating is somewhat higher than the one in Poland, meaning that on average Hungarian farmers have a slightly more positive impression on the adaptation to changes in national regulations. When considering the proportion of farmers who did not have any opinion on the issue, we can see that Spanish farmers still steadily express their opinions while around 3% of Hungarian and about 7% of Polish and Dutch farmers indicated no opinion preceded by Germany with nearly 13%.

If we take an alternative approach (tables not presented) and compare the average ratings of the support from the EU and the national institutional environment in each selected member state, it can be observed that in Hungary, Poland and the Netherlands farmers consider the EU institutional environment more supportive than the national one. In Spain and Germany there was no statistically significant differences revealed between these issues, so neither one is preferred to the other. When looking at the ease of adaptation to changes in the EU and the national regulations, we can see that according to Hungarian and Polish farmers it is somewhat easier to adapt to the changes in the EU regulations than to those on a national level. In case of the other three member states no difference could be revealed.

## 5.4 Discussion and conclusions

When considering factors that affect farming, it became clear that farmers in the studied member states and also in the different economic size and farming activity groups perceive weather and natural disasters, and price volatility the elements having the largest effects on their farming. It strongly corresponds to the finding that climatic and market conditions were the primary reasons of critical situations experienced by farmers. Based on the relevant results, production and related revenues are the most sensitive to crises in Spain and Hungary, both in crop and livestock production, while the other member states, especially Poland are quite secure relative to these two member states in this respect.

The range of instruments applied by farmers to manage risks related to agriculture show that specific crop and livestock insurance is widespread mainly in the established member states while property insurance has an important role in both the established and new member states. Although, the established member states have well developed financial markets, a high proportion of farmers in all selected member states tries to avoid using credit and thus taking on new liabilities. A widespread way to secure sufficient financial resources is to hold financial reserves, especially in the new member states and Germany. This method is reasonable in new member states where the majority of farmers perceive the adequacy of access to credit in a less positive way than those in the established member states.

When future use of risk management instruments is considered, it becomes quite clear that the majority of farmers finds the instruments they currently use as adequate for use in their farming to offset risks, so they are going to continue applying them. The same applies to the economic size approach, where majority of farmers would apply the currently used methods also in the future, although it has to be noted that with the increase of economic size of farms the farmers are more interested in hedging (futures and options), which is a sophisticated method of risk management.

A special way of dealing with risks related to agriculture is to find work outside farming and generate revenues additional to farming and thus volatility of farm incomes can be offset by other income sources. In the established member states a higher proportion of farmers have off-farm revenues (except the Netherlands where farming provides sufficient earnings) than in the new member states where there are no job opportunities outside farming in rural areas. However, where off-farm revenue is available it is generated continuously throughout the year in the majority of cases, both in established and new member states.

An important aspect of risks surrounding farmers is to see how they perceive the EU and national institutions and regulatory changes related to farming. We can conclude that farmers do not attribute high risk to these institutional settings and the adaptation process to changes in regulations, however there is still space to improve farmers' acceptance and opinion both on the EU and national levels.

The most important conclusion of the study presented here is that, although high similarity was expected between pairs of member states like Germany-Netherlands and Hungary-Poland in terms of farmers' risk perception and management strategies, it has to be clearly seen that strong similarity can be concluded only at the level of individual, highly specific issues, because at the general level there are many differences among these given member states. The case of Spain is an outstanding example for how complex the network of similarities and differences can be as it is very close to other member states in some respects, while in the case of other issues it is very different from the same member states. It has to be strongly highlighted that this survey has covered only five member states out of the twenty-seven so there may be many more combinations of similarities and differences among member states besides the ones revealed. Although in several cases member states with similar conditions (economic, social, natural, etc.) may show more features in common but sometimes surprising differences could be revealed (e.g. Hungary and Poland). That is why, on the European Union level, the adequate answer for such challenges could lie in the establishment of a flexible risk management policy framework which could be well customised based on the specific needs of individual member states while meeting universal guidelines across the Union.

# Chapter 6

# Economic impact of prospective risk management instruments under alternative policy scenarios

*Marcel A.P.M van Asseldonk, Miranda P.M. Meuwissen and Ruud B.M. Huirne*

## 6.1 Objectives

The projected World Trade Organisation (WTO) and Common Agricultural Policy (CAP) policy scenarios have created an awareness amongst farmers and policy makers of the need to quantify the altered risk exposure, some of which being catastrophic and disruptive, and study the scope for better risk management opportunities. In this chapter the objective is to quantify the economic impact policy scenarios in conjunction with a set of prospective risk management instruments for the European Union. In order to perform these analyses a whole-farm model is developed which provides insight into the impact of (new) instruments on farm income volatility and farm crisis risk. Building on the Monte Carlo simulation model developed in Chapter 3 the impact of instruments is determined.

## 6.2 Introduction

A considerable amount of general work has been done on issues such as on-farm risk management and the use of some well-established risk-sharing instruments such as commodity futures and conventional indemnity insurance. There are, however, three main limitations to this accumulated body of knowledge. First, new instruments are being developed apace, creating a need of analysing their worth for the specific farming systems under consideration. Second, a large proportion of the past work on financial instruments for risk management has been done using partial analysis, i.e. assessing the merits of using each particular risk/risk instrument alone. However, such partial analyses may be flawed, perhaps misleading, especially (but not solely) where the instruments or strategies analysed are strongly negatively correlated with other parts of the farm firm. Third, the analyses were very general in nature and not tailored to individual (farm) production circumstances. It is hardly recognised that the farm manager himself has a potential to influence and improve his own risk situation: through on-farm risk management strategies (such as technology choice, risk prevention, flexibility and diversification) and through sharing risks with others (such as hedging, contract farming and insurance).

Consider a newly introduced form of insurance that indemnifies farmer policy-holders against low income from a particular crop. By taking up such a policy a farmer will normally be accepting a small reduction of expected net returns from the crop (mainly due to the premium paid), but is guarding against very unfavourable outcomes. On the face of it, it seems that only a risk-averse farmer would consider buying such a policy and the decision would depend on the cost of the premium relative to the benefit perceived from the reduction in down-side risk, as well as on the degree of risk aversion. With information about the farmer's attitude to risk and about the probability distribution of returns from the crop, a choice to insure or not can apparently be rationalised. However, that is not the whole story. First, our farmer has now 'locked out' one source of risk of a bad overall outcome, so he or she can now be rather more adventurous in making other risky decisions about next year's production. These more adventurous decisions are likely to increase expected returns at the cost of some increase in the risk of a adverse result. Moreover, almost by definition, an insurance instrument is a risky prospect that is negatively correlated with the risk insured. Yet in farming there is often a positive correlation between risky production prospects – most crop yields are positively correlated, and so too are many product prices. A negative correlation of the insurance with the returns from crops and livestock might mean that, by purchasing the insurance, the farmer can in effect 'trade away' not only some of the risk in the specific production insured, but also some part of the risk in other production activities. Moreover, with this additional risk reduction, he or she is still better placed to be more adventurous in choice of production options. For example, the availability of a new risk management instrument may allow the farmer to borrow more to invest in improving productivity.

In general, it will be impossible to say whether the net effect of the introduction of a new risk management instrument will increase or reduce either the mean or the variance of net returns. It depends on how the interactions with other risks on the farm and with other risk management instruments work out. All we can be sure of is that, if the decisions are taken rationally, the farmer's utility should not go down and would normally remain the same only if he or she found the new instrument unattractive. Thus the merit of adding any risky prospect into an existing farm business cannot be assessed without considering the potential impact on the risk-efficiency of net returns from the whole portfolio of farm-specific risky prospects (including any off-farm investments or income-earning ventures). This is true whether the added prospect is in the form of a new production activity, a new policy, or a new risk management instrument. And, in making an evaluation, it is necessary to take account of the stochastic dependencies, such as the correlations, between the new activity and the existing ones (Hardaker *et al.*, 2004). In order to perform these analyses a whole-farm model is developed which provides insight into the impact of (new) instruments on farm income volatility.

## 6.3 Method and outline

### *6.3.1 Whole-farm analysis*

Risky decision problems are often handled by means of portfolio optimisation. Portfolio analysis for farm planning requires the inclusion of the normal range of risky production activities and should comprise the probability distribution of per unit net revenue for each activity and the stochastic dependencies between those activities.

Markowitz (1959) as well as Freund (1956) showed that quadratic risk programming (QRP) can be used to maximise the expected income of a risk-averse decision-maker subject to a set of resource and other constraints including a parametric constraint on the variance of income. The model can also be formulated to minimise the variance subject to a parametric constraint on expected income, or to expected constant absolute risk aversion (CARA) utility maximisation with parametric variation in absolute risk aversion. All three should give identical solutions.

QRP restrictively uses the first two moments (i.e. mean and variance) of each risky activity and the first co-moment (i.e. covariance) between the risky activities. The obtained optimal portfolio with respect to income or wealth is usually held to be a reasonable approximation provided that the distribution of income or wealth is not very skewed. Note that the activity per unit net revenues may not have to be normal distributed for the distribution of farm income or wealth to be more or less normal. Under some particular assumptions, it is exact, e.g. when the distribution of income is normal and the utility function is negative exponential (Freund, 1956) or when the utility function is quadratic (Anderson *et al.*, 1977). The risky alternatives can subsequently be ranked by applying the stochastic efficiency with respect to a function method (SERF). This method allows comparing the alternatives in terms of certainty equivalents (CE) over the range of risk aversion of interest. SERF works by identifying utility-efficient alternatives for ranges of risk attitudes and can be applied to any utility function.

As an alternative, a non-parametric risk-programming method is free of distribution assumptions and includes the joint distribution by means of so-called 'states of nature' (i.e., specific combinations and probabilities of possible outcomes). Utility-efficient programming (UEP) is one of the non-parametric methods applied in farm portfolio analysis. The UEP is formulated as follows:

$$\max E[U] = pU(z,r),\ r \text{ varied}, \tag{1}$$

subject to:

$$Ax \leq b \tag{2}$$

$$Cx - Iz = f \tag{3}$$

$$x \geq 0 \tag{4}$$

where: $E[U]$ is expected utility, $p$ is vector of probabilities for states of nature (often assumed equi-probable), $U(z,r)$ is a vector of utilities of net income where the utility function is defined for a measure of risk aversion, $r$, $A$ is a matrix of technical coefficient, is a vector of activity levels, $b$ is a vector of resource stocks, $C$ is a matrix of GMs for $S$ states of nature, $I$ is a identity matrix, $z$ is a vector of net incomes for each state of nature $S$, $f$ is a vector of fixed costs.

Berg and Starp (2006) use shortfall models to optimise the production plan. They argue that analysing the trade-offs as loss expectation and mean profit is intuitive more appealing than presenting the trade-offs as risk aversion and CE. The lower part of the distribution is considered and account for the downside-risk and are referred to as lower partial moments (LPM). LPM1(π) denotes the expected value of shortfalls multiplied by the probability of the occurrence of below target returns (π). Thus, it accounts for the probability as well as for the magnitude of shortfalls. A target return of zero (z=0) implies that the expected value of negative outcomes is used as risk measure. The model is set up as to compute a risk efficient frontier in the way that the expected profit enters the objective function while the risk measure is considered as a constraint; LPM1(0)≤c, where c is parameterised in order to compute the efficient frontier.

### *6.3.2 Risk management tools*

The described models (QRP, UEP and shortfall) can be augmented to optimise the portfolio of crops grown in the coming year, including options to insure a shortfall of the long-term average (in case of yield or revenue insurance) or an insurance scheme based on an index.

The objective of yield insurance is to reduce the fluctuations in income caused by yield variations. Yield insurance indemnifies any insured farmer in any year in which yield falls below a specified level (coverage level). This strike level is defined as a farm-specific percentage of the expected yield per hectare (Halcrow, 1949). Crop revenue in case of yield insurance equals:

$$IR_{qn} = R_{qn} - IP_{qn} + PInd_{qn} \cdot (\overline{Y}_{qn} \cdot C - Y_{qn}),\ \text{if } Y_{qn} < \overline{Y}_{qn} \cdot C \qquad (5)$$

where $IR_{qn}$ is revenue of crop $q$ in case if insurance applied on farm $n$; $R_{qn}$ is observed revenue of crop $q$ on farm $n$, which is calculated as: $R_{qn}=Y_{qn}P_{qn}$, where $Y_{qn}$ is observed yield of crop $q$ on farm $n$ and $P_{qn}$ is observed price of crop $q$ on farm $n$; $IP_{qn}$ is insurance premium of crop $q$ on farm $n$; $PInd_{qn}$ is the indemnity price of crop $q$ on farm $n$ (which can be established by the farmer or can be nominated by the insurance company; in each case always at the beginning of the contract year), $Y_{qn}$ is average yield of crop $q$ on farm $n$; and $C$ is coverage percentage level.

Insuring revenue of a given crop implies insuring the product of price and yield of that crop. For revenue insurance it is important to consider the joint distribution of prices and yields. Farm total revenue from crops with crop revenue insurance equals (Kaylen *et al.*, 1989):

$$IR_{qn} = R_{qn} - IP_{qn} + (\overline{R}_{qn} \cdot C - R_{qn}),\ \text{if } R_{qn} < \overline{R}_{qn} \cdot C \qquad (6)$$

where $\overline{R}_{qn}$ is average revenue of crop $q$ on farm $n$ that is calculated as:

$\overline{R}_{n} = \overline{Y}_{n} \cdot \overline{P}_{n}$.

Besides indemnity-based insurance schemes, also index-based insurance are of interest in the current study. In this insurance scheme, the premiums and payouts are based on the weather records of the locality in which the insurance is sold (Halcrow, 1949). Payouts to a farmer are triggered if weather, in terms of some measurable criterion, is below the certain limits of tolerance. Weather index-based insurance would be adapted more easily to an area in which one or two weather factors such as precipitation and temperature are generally limiting and are highly significant in projecting crop yields (Halcrow, 1949). So any applied index only accounts for a certain amount of the total risk (i.e. basis risk). Basis risk refers to the inadequate stochastic dependency between the actual weather risk exposure of the buyer and the outcome of the weather underlying the hedging instrument.

In terms of risk programming, index insurance products can be incorporated by assuming that only a certain percentage of observed adverse years are eligible for compensation reflecting the associated basis risk.

### *6.3.3 Computations and assumptions*

Since the future is uncertain whole-farm plans are depended on the assumed CAP and WTO policy scenarios. These scenarios have their implications for price, yield, farm income and allowed forms of income stabilising tools. The formulation of future CAP scenarios was derived on basis of three main components: (1) progress in the WTO negations; (2) changing public expectations and increasing scrutiny as regards the role and the efficiency of the CAP; and (3) the accelerated debate on the EU budget. Given three time perspectives (current, 2013 and 2018) and alternative assumptions about CAP we evaluate six scenarios (Table 6.1). Three potential risk management tools were evaluated, namely: (1) yield insurance per crop; (2) revenue insurance per crop; and (3) index insurance per crop.

Different farming systems per member state were selected for in-depth analysis. Specialised cereals, oilseed and protein crop farms (FADN typology 13) were included in the analysis for Hungary, Spain and Poland, whereas general field cropping farms (FADN typology 14) were considered for Germany and the Netherlands. Since average farm size differs considerably between those member states this was taken into account (Appendix Table D.1).

Building on the Monte Carlo simulation model (Majewski *et al.*, 2007) farmers' choices are being modelled given scenarios of CAP regulations and potential risk

**Table 6.1.** *Policy scenarios.*

| Year | Scenario | Description |
|---|---|---|
| 2004 | Base | Historic reference |
| 2013 | ML13 | Luxembourg 2003 policy implemented (sugar reform), no substantive policy changes, modulation -10% |
| 2018 | LikA18 | Higher support level, full de-coupling, mandatory modulation -10% |
| | LikB18 | Lower support level, full de-coupling, ceiling 100,000 euro, mandatory modulation -20% |
| | Lib18 | Non-tariff market protection measures removed, no direct payments |
| | Pro18 | Return to 'pre-CAP' reform type of policy - stronger market protection |

management tools. By means of the Latin Hypercube procedure 100 states of nature were sampled, entailing yields, farm gate prices as well as variable and fixed costs of production. This procedure was taken in favour of Monte Carlo simulation (MCS) because it divides the distribution on the equal number of intervals so that tails with a downside risk and upside potential are taken into account (Richardson *et al.*, 2006). Contrary, MCS randomly selects points, so that the tails can be underestimated even with higher number of iterations. Subsequently, the impact of the CAP scenario on the portfolio decisions made were derived under the assumption of risk neutral behaviour (linear programming model) as well as moderate risk aversion behaviour (risk programming model). Three alternative risk programming models were applied with and without hypothetical insurance contracts (QRP, UEP and shortfall), while output reported are the expected farm income, the probability of a negative farm income, coefficient of variability (CV) and optimal farm plans. The models were solved using MATLAB / FMINCON allowing a constraint function that has both equality and inequality elements.

## 6.4 Results

### *6.4.1 Whole-farm optimisation results*

In general, optimal production plans are very sensitive towards constraints as well as variable costs imposed. A key dilemma is how to handle the constraints and variable costs for the category of aggregated crops, hence referred to as 'residual crops'. It is observed from the FADN records that the proportion of land allocated for it differs between the analysed farming systems. While the proportion of land cultivated with other crops is trivial for a typical farm in Poland (1% of 50 ha), it is substantial for the other four typical farms (13% of 52 ha in Hungary, 20% of 66 ha in Germany, 28% of 80 ha in Spain, and 39% of 41 ha in the Netherlands). Cultivated crops in this category are divers with respect to revenues received and variable cost paid. It can be permanent grassland, but also an orchard in Poland, or olive trees in Spain, or more capital intensive arable or vegetable crops in Germany and the Netherlands. Omitting this category of crops will affect the optimal farms plans and expected farm income considerable and becomes untrustworthy to reflect the options available to these typical farming systems. In the optimisation procedure this category is treated as a 'fixed' constraint imposing that the observed (residual) proportion of area is allocated for it. Note that the residual revenues remains stochastic as generated by the Monte Carlo simulation model and the joint distribution still captures the (co-) variability between the revenues of residual crops with yields and prices of the main crops. Given this fuzzy defined category it is infeasible to derive variable costs per hectare from normative sources. Therefore variable costs per hectare observed are assigned. The selected typical arable farming systems also constitute livestock activities (i.e., dairy, cattle, sheep, pigs and poultry) at a marginally level and are therefore ignored.

Variable costs (Appendix Table D.2) and constraints (Appendix Table D.3) for the main crops are determined individually for each farming system modelled. Crops considered are wheat, rye, barley, oats, triticale, maize, other cereals, potatoes, sugar beet, rapeseed and sunflowers. However, not all crops are common in certain regions. For example, rapeseed is hardly cultivated in Spain while sunflowers are, whereas the opposite holds for the Netherlands. Specificity of the farming system (e.g., possibility of irrigation, quality of soils which is not shown in FADN data, but observed crop selection provides some information) as well as normative sources of information are taken into account. For example variable costs vary a lot in Spain, depending on whether or not crops are irrigated. If they are, yields are high and production is very intensive and some crops, can only be cultivated in rain-fed regions (in the crops under investigation maize, sugar beet and potatoes require irrigation). The expected variable costs are calculated as weighted mean values for farms with different quality soils and intensity of production (Appendix Table E.2). Also constraints of feasible cropping plans are differentiated for specific farm types according to the observed structure. If cropping plans are solely restricted on agronomic insights obtained under optimal conditions the models may explode with amounts exceeding what the market realistically may absorb. For example, if potatoes are allowed to be grown more or less freely of the area available an excessive amount will be grown because of its high profitability.

The joint future performance distribution were derived from a Monte Carlo simulation model which depended strongly on the assumptions made as well as the quality of the entry data, largely coming from the FADN database. Differences between simulation results presented in one of the previous chapters and optimisation results depicted in Table 6.2 originate by alternative production plans and alternative variable and fixed costs assumptions.

The pattern of changes in the level of expected farm income across scenarios is similar for the five case farms under investigation. On the long run expected farm incomes increase under protectionist policy (Pro18) but are depressed if liberalisation is assumed (Lib18). The impacts of alternatively policy scenarios on the optimal farm plan (i.e. level of activities) were not substantial. The allotted acreage in the farm plan of cash crops such as sugar beet and potato, which were the most profitable cropping activities considered, corresponded to the maximum proportion allowed. This is to say when decisions are made assuming risk neutrality whereby farmer are not willing to forego a part of the expected income in order to avoid the risks associated with the cultivation of these risky cash crops. As a

**Table 6.2.** *Linear programming results (risk neutral).*

| | Farm size[1] | Farming system[2] | | Scenarios | | | | | |
|---|---|---|---|---|---|---|---|---|---|
| | | | | Base | ML 13 | LikA18 | LikB18 | Lib18 | Pro18 |
| Germany | ≥40-<100 | 14 | E (euro) | 30,540 | 10,803 | 26,146 | 20,917 | 12,863 | 50,150 |
| | | | CV (%) | 109 | 395 | 210 | 292 | 501 | 104 |
| | | | P<0 (%) | 17 | 47 | 37 | 44 | 49 | 17 |
| Hungary | ≥8-<16 | 13 | E (euro) | 25,425 | 28,344 | 29,848 | 29,360 | 19,543 | 34,523 |
| | | | CV (%) | 57 | 60 | 67 | 72 | 106 | 54 |
| | | | P<0 (%) | 0 | 0 | 0 | 0 | 14 | 0 |
| Netherlands | ≥40-<100 | 14 | E (euro) | 9,521 | -10,583 | -6,418 | -8,488 | -9,288 | -7,666 |
| | | | CV (%) | 529 | -492 | -946 | -716 | -654 | -833 |
| | | | P<0 (%) | 50 | 61 | 60 | 60 | 60 | 60 |
| Poland | ≥8-<16 | 13 | E (euro) | 18,567 | 21,051 | 21,553 | 21,117 | 13,871 | 23,604 |
| | | | CV (%) | 24 | 24 | 25 | 26 | 39 | 25 |
| | | | P<0 (%) | 0 | 0 | 0 | 0 | 0 | 0 |
| Spain | ≥16-<40 | 13 | E (euro) | 18,411 | 13,716 | 14,881 | 12,669 | 2,712 | 19,150 |
| | | | CV (%) | 42 | 65 | 65 | 76 | 357 | 54 |
| | | | P<0 (%) | 0 | 5 | 5 | 12 | 39 | 3 |

[1]European Size Unit.

[2]Farming systems: 13=specialised cereals, oilseed and protein farm; 14=specialised arable farm.

result, general field cropping farming systems (FADN typology 14) which farm plan can constitute a relative large proportion of these cash crops have a more volatile farm income than specialised cereals, oilseed and protein farms (FADN typology 13). The coefficient of variability as well as the probability of a negative farm income are for the two general field cropping case farms considerable. Both effects originate from volatile crop revenues in conjunction with relative high cost causing a relative low expected farm income.

To evaluate the impact of insurance within an optimal farm portfolio context three additional optimisations were run. In each optimisation only insured activities were considered, being either yield, revenue or index insurance. Note that in the current analysis the strike level is set at 80% of the mean, implying a deductible of 20%. The risk reducing impact of the three insurance schemes under investigation in terms of CV is presented in Table 6.3. It was assumed that the chance of payments via the index insurance was 75% if actual losses were incurred (i.e., basis risk). This basis risk was captured in the model empirically; by means of a randomisation procedure the probability of payouts was introduced.

The relevance of insurance contracts in terms of its risk reducing impact might be derived by comparing the CV's obtained with and without insurance. For all case farms and scenarios the revenue-coverage contract was most effective, and reduced CV on average by about 22%, followed by yield insurance (-13%) and index insurance (-5%). The efficacy was more or less independent from the scenario considered. Also the impacts of (alternative) insurance contract on the optimal farm plan were not substantial. Obtained results are counterintuitive if efficacy of insurance – being either yield, revenue or index insurance – is expressed in terms of its risk reducing impact on the probability of negative farm income (not reported). The probability of a negative farm income hardly reduces and sometimes increases if crops are insured. These results can be explained by the fact that this parameter captures the efficacy partially. Extreme negative yields and revenues are indemnified, but in case of low expected incomes relative to its variability already moderate adverse years will generate negative farm incomes because of the premiums to be paid. In general, from the results it can be seen that the net effect of the introduction of a new risk-management instrument will affect the variability of farm incomes, as theory suggests. Of course, the efficacy can be expressed in alternative means. All we can be sure of is that, if the decisions are taken rationally, the farmer's utility should not go down and would normally remain the same only if he found the new instrument unattractive.

The pure premiums, also referred to as the expected claim cost or actuarially fair premium, for each type of insurance given a particular farming system are

**Table 6.3.** *Impact of alternative insurance options (CV,%).*

| | Farm size[1] | Farming system[2] | Insurance option | Scenarios | | | | | |
|---|---|---|---|---|---|---|---|---|---|
| | | | | Base | ML 13 | LikA18 | LikB18 | Lib18 | Pro18 |
| Germany | ≥40-<100 | 14 | No | 109 | 395 | 210 | 292 | 501 | 104 |
| | | | Yield | 100 | 350 | 188 | 267 | -46 | 93 |
| | | | Revenue | 89 | 312 | 166 | 227 | 390 | 82 |
| | | | Index | 107 | 378 | 201 | 309 | 488 | 99 |
| Hungary | ≥8-<16 | 13 | No | 57 | 60 | 67 | 72 | 106 | 54 |
| | | | Yield | 51 | 54 | 59 | 65 | 99 | 46 |
| | | | Revenue | 45 | 48 | 53 | 58 | 86 | 42 |
| | | | Index | 54 | 58 | 64 | 69 | 103 | 50 |
| Netherlands | ≥40-<100 | 14 | No | 529 | -492 | -946 | -716 | -654 | -833 |
| | | | Yield | 495 | -461 | -882 | -670 | -613 | -780 |
| | | | Revenue | 406 | -379 | -725 | -551 | -504 | -640 |
| | | | Index | 499 | -477 | -926 | -699 | -638 | -815 |
| Poland | ≥8-<16 | 13 | No | 24 | 24 | 25 | 26 | 39 | 25 |
| | | | Yield | 20 | 20 | 21 | 22 | 34 | 20 |
| | | | Revenue | 20 | 20 | 21 | 22 | 32 | 20 |
| | | | Index | 23 | 22 | 23 | 24 | 36 | 22 |
| Spain | ≥16-<40 | 13 | No | 42 | 65 | 65 | 76 | 357 | 54 |
| | | | Yield | 21 | 34 | 20 | 32 | 28 | 50 |
| | | | Revenue | 21 | 32 | 20 | 32 | 24 | 50 |
| | | | Index | 23 | 36 | 22 | 37 | 30 | 58 |

[1]European Size Unit.

[2]Farming systems: 13=specialised cereals, oilseed and protein farm; 14=specialised arable farm.

presented in Table 6.4. Note that converting the pure premium into a gross rate requires the addition of the loading, which is intended to cover transaction costs and allowance for contingencies and profit (this aspect will be elaborated on in the section describing the budgetary implications).

Levels of pure premiums per hectare differed between case farms and were affected by the alternatively policy scenarios. On the long run expected premiums increased under protectionist policies as well as more liberal policies. For German and Dutch case farms premiums charged for the revenue-coverage contract exceeded those for yield insurance and index insurance. Revenue insurance premiums on general field cropping farming systems, with more volatile cash crops (i.e., price variation), were higher than those on specialised cereals, oilseed and protein farms (i.e. relative low variation).

### 6.4.2 Sensitivity analysis whole-farm optimisation

In this section, what-if analyses are carried out with respect to (1) the impact of risk aversion; (2) the impact of subsidy; and (3) the variability based on three-year averages.

In Figure 6.1 the results are shown whereby the bars represent the outcomes under the assumption of a moderate risk averse decision maker while lines represents deviations from the results obtained under risk neutrality outlined previously (the procedure to capture moderate risk aversion into risk programming is described by Hardaker *et al.*, 2004). According to the model, the trade off between risk and profit was at a fairly low rate given moderate risk-averse decision makers. The optimal expected farm incomes were slightly lower under risk aversion than under risk neutrality. Again some counterintuitive results were obtained if the

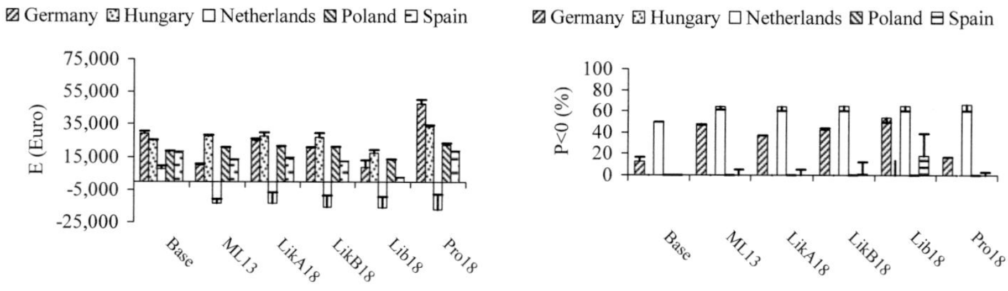


**Figure 6.1.** *Whole-farm results under risk averse decision making (versus default results). Bars represent outcomes of the sensitivity analysis while lines represent deviations from default assumptions.*

**Table 6.4.** *Premium of alternative insurance options (Euro per hectare).*

| | Farm size[1] | Farming system[2] | Insurance option | Scenarios | | | | | |
|---|---|---|---|---|---|---|---|---|---|
| | | | | Base | ML 13 | LikA18 | LikB18 | Lib18 | Pro18 |
| Germany | ≥40-<100 | 14 | Yield | 84 | 103 | 118 | 100 | 47 | 118 |
| | | | Revenue | 171 | 207 | 265 | 288 | 310 | 237 |
| | | | Index | 52 | 65 | 73 | 58 | 61 | 73 |
| Hungary | ≥8-<16 | 13 | Yield | 44 | 55 | 61 | 63 | 61 | 63 |
| | | | Revenue | 79 | 91 | 104 | 109 | 114 | 92 |
| | | | Index | 22 | 27 | 31 | 31 | 30 | 31 |
| Netherlands | ≥40-<100 | 14 | Yield | 76 | 85 | 87 | 89 | 89 | 89 |
| | | | Revenue | 324 | 345 | 391 | 404 | 407 | 415 |
| | | | Index | 49 | 54 | 56 | 57 | 57 | 57 |
| Poland | ≥8-<16 | 13 | Yield | 164 | 187 | 201 | 199 | 180 | 201 |
| | | | Revenue | 14 | 15 | 16 | 17 | 18 | 17 |
| | | | Index | 71 | 81 | 86 | 87 | 86 | 87 |
| Spain | ≥16-<40 | 13 | Yield | 29 | 34 | 38 | 23 | 38 | 38 |
| | | | Revenue | 31 | 34 | 37 | 34 | 38 | 41 |
| | | | Index | 15 | 17 | 19 | 18 | 19 | 19 |

[1]European Size Unit.

[2]Farming systems: 13=specialised cereals, oilseed and protein farm; 14=specialised arable farm.

impact of risk aversion is expressed in terms of its risk reducing effect on the probability of negative farm income. This can be explained by the fact that the expected utility is maximised and not the probability of a negative farm income is minimised. Comparing the E,V results with the UEP results showed that there are few differences between the two and the differences which do occur are mainly trivial. In general, it was observed, that if a farmer was more risk-averse, he was more prone to choose a production plan comprising more less-profitable lower-variance crops (wheat instead of potato) compared to the optimal plan achieved.

The amount of subsidy received differs per scenario. In Figure 6.2 the results are shown whereby the bars represent the outcomes under the assumption that subsidy is absent while lines represents deviations from the results obtained with subsidies outlined previously. Without subsidy expected farm incomes decreased sharply, becoming in certain situations even negative. The opposite holds for the probability of negative incomes. Differences were more substantial under protectionist policies than under liberal policies. The impacts of alternatively policy scenarios on the optimal farm plan were limited since future subsidy schemes were not activity related (decoupled).

In Figure 6.3 the results are shown whereby the bars represent three-year average farm incomes. Expected farm incomes are not altered, but the probability of a negative income in three years decreases. The absence of autocorrelation implied that a series of farm incomes values becomes less volatile. For the German and Dutch case farms the probability of negative outcomes reduced substantial. Negative outcomes dropped in the range of 15 up to 25 percent points depending on the scenario. The multi-year impact for the other case farm was negligible since the annual level was already low. However, a profound risk reduction was measured for all case farms if it is measured in for example CV levels.

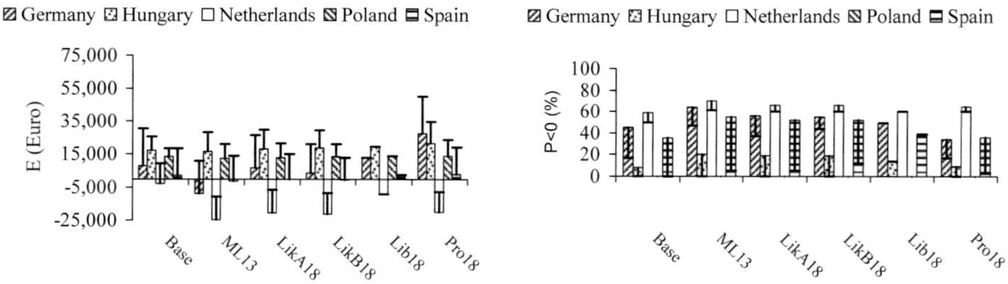


**Figure 6.2.** *Whole-farm results without subsidy (versus default results). Bars represent outcomes of the sensitivity analysis while lines represent deviations from default assumptions.*

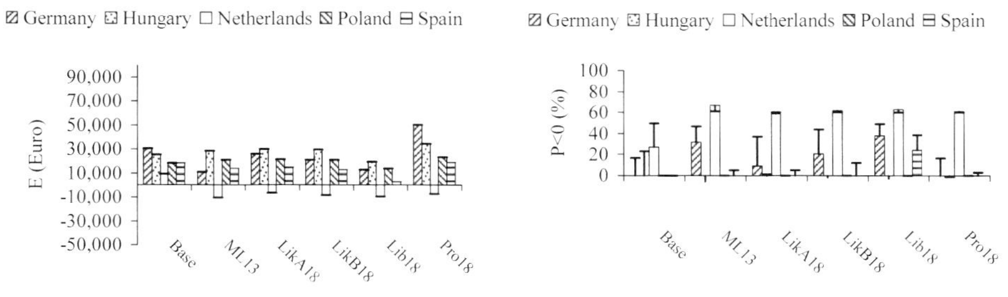


**Figure 6.3.** *Whole-farm results based on three-year average (versus default results). Bars represent outcomes of the sensitivity analysis while lines represent deviations from default assumptions.*

Making deposits to saving accounts in favorable years and making withdrawals in adverse years is a form of risk protection strategy. Note that the efficacy of these so-called income stabilisation accounts may be limited due to the lack of adequate account balances and buildup of balances beyond the level required for risk management (Dismukes and Durst, 2006).

### *6.4.3 Budgetary implications*

Budgetary expenses are depended on the assumed CAP and WTO policy scenarios in two ways. First, the amount of subsidy received by farmers differs per scenario but have in common that payments are decoupled. Second, the public budget is also affected if premiums are partly subsidised and/or governments provide protection in case of contingencies (i.e., reinsurance). Both forms of public support are interconnected since the considered alternative scenarios affect price and farm income volatility. The scenario dependent pure premiums were determined in the previous paragraph (given four case farms and three alternative insurance schemes). However, to convert the pure premium into a gross rate requires the addition of the loading.

It not straightforward and thus arbitrary to derive loadings since, among others, the (re)insurance market is not transparent and the level of loading depends on the market cycle (soft versus hard). Also the requirements differ between market based insurance schemes versus state support schemes with respect to amounts of operating capital. As a benchmark, market based agricultural hail insurance schemes in Europe operate with a long-term average loss ratio of about 55%, which is the percentage of the rate that is intended to cover losses (Van Asseldonk *et al.*, 2006). However, yield insurance, revenue insurance and index insurance are less conventional insurance schemes with distinctive loadings. Therefore to mirror the (current) market situation the level of loading need to be addressed per insurance scheme. Not that the global market experience is dominated by the North American (USA/Canada) (re-)insurance market. Outside North-America crop revenue covers does almost not exist and crop index insurance has – at least until today – no significant commercial importance.

Loading for yield insurance comprise 27.5% up to 32.5% for transaction costs. In addition, costs associated with capital requirements need to be accounted for, and range from 50% up to 100% of gross premium income (GPI). Capital requirements vary according to data availability and state support (USA with high state support have low capital requirements while countries without state support have higher capital requirements). Finally, 15% to 18% return on equity (ROE) is common. Total loadings are therefore approximately 40% to 45%; implying a required long-term target loss ratio of 55% or better. Capital requirements for crop revenue insurance are approximately 50% to 75% of GPI in North-America. Assuming other loading factors identical to that of yield insurance results in a total loading of approximately 37.5% up to 42.5% and thus a long-term target loss ratio of minimum 57.5%. Loading for a weather index insurance comprise 17.5%-25.0% for transaction cost, capital requirements are 75% to 100% of GPI. Total loadings are therefore approximately 30% to 35% (and a long-term target loss ratio required of 65% or more).

Various forms of subsidised multi-peril crop and farm income insurance exist in a number of countries, such as the USA and Canada. These comprehensive public-private schemes are available in some EU member states. However, it is unclear which forms of public-private stabilising tools are allowed at which levels in the future scenarios. To address the implications for budgetary expenses we depict in Figure 6.4 the payments involved with insurance. Note that budgetary consequences stemming from insurance depend on the level of granted premium subsidy. A loading of 45%, 42.5% and 35% is added to the expected indemnity payments for yield, revenue and index insurance respectively.

Total payments per hectare differed between case farms and were affected by the alternatively policy scenarios. Gross premium payments on general field cropping farming systems (Figures 6.4A and 6.4C), with more volatile cash crops, exceeded those on specialised cereals, oilseed and protein farms (Figures 6.4B, 6.4D and 6.4E). Expected gross premiums under more liberal policies were more substantial than under protectionist policies.

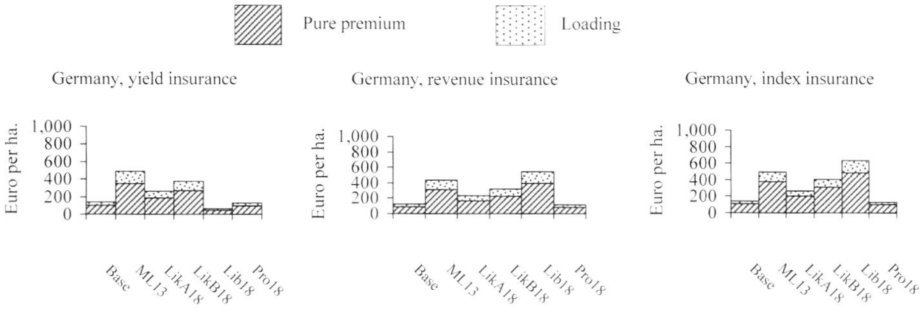


**Figure 6.4A.** *Amount of premium paid per insurance scheme (pure premium and loading) given Germany case farm.*

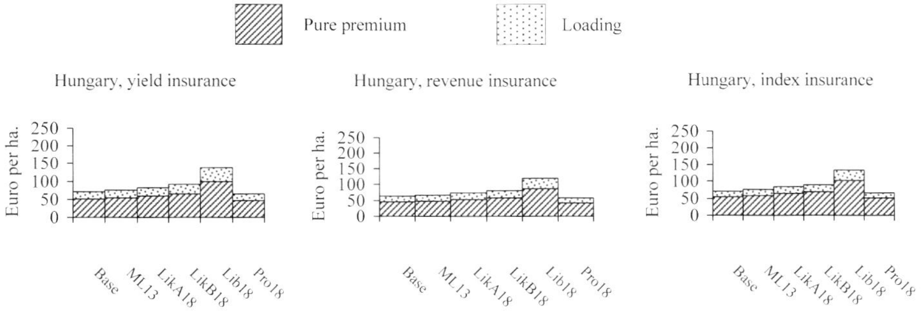


**Figure 6.4B.** *Amount of premium paid per insurance scheme (pure premium and loading) given Hungarian case farm.*

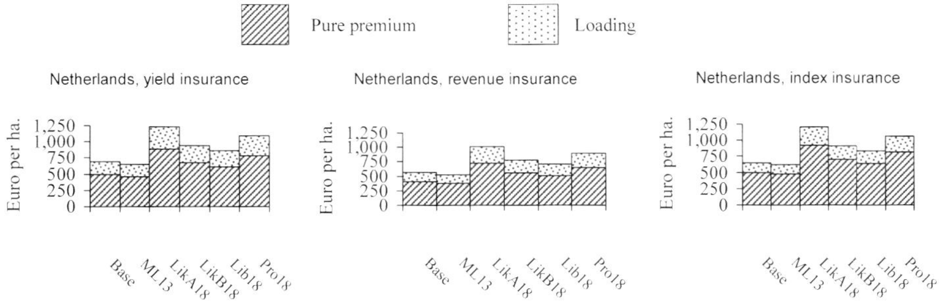


**Figure 6.4C.** *Amount of premium paid per insurance scheme (pure premium and loading) given Dutch case farm.*

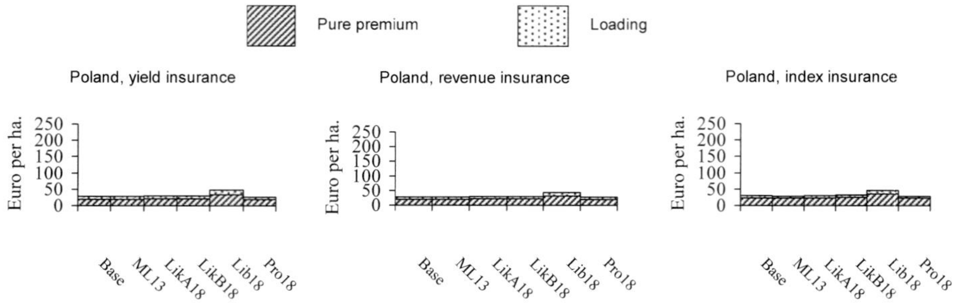


**Figure 6.4D.** *Amount of premium paid per insurance scheme (pure premium and loading) given Polish case farm.*

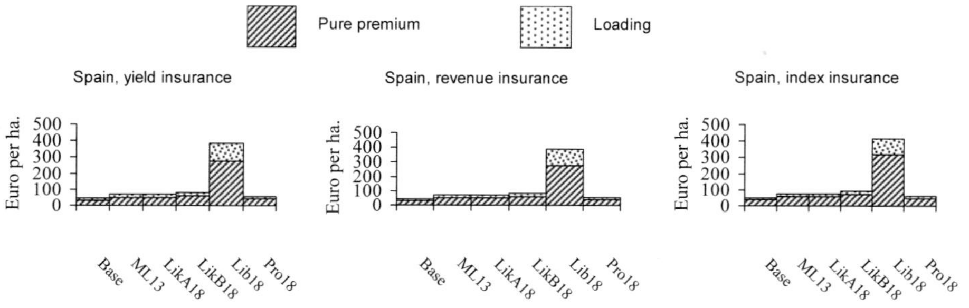


**Figure 6.4E.** *Amount of premium paid per insurance scheme (pure premium and loading) given Spanish case farm.*

## 6.5 Main conclusions and discussion

The goal of this chapter was to investigate the economic impact of future policy scenarios in conjunction with a set of prospective risk management instruments for the European Union. The following five main conclusions were derived by analysing five case farms, while their general implications and applicability are discussed subsequently:

1. Farm risk exposure differed between the assumed future scenarios substantially. The pattern of changes in the level of expected farm income across scenarios is similar for the five case farms under investigation. On the long run expected farm incomes increase under more protectionist policies but are depressed if liberalisation is assumed.

2. The impacts of alternatively policy scenarios on the optimal farm plan were not substantial. The optimal farm plan of general field cropping farming systems as well as specialised cereals, oilseed and protein farms is marginally altered. The amount of cash crops cultivated – which are characterised by higher but more volatile outcomes – is more affected by agronomic constraints rather than future policy scenarios.
3. Diversification as a risk management tool has its limitations. The analysis of the case-specific trade-off between the expected gross margins and risk provided an indication of the efficiency of farm diversification. This is to say when decisions are made assuming risk neutrality or moderate risk aversion whereby farmers are not willing to forego a substantial part of the expected income in order to avoid the risks associated with the cultivation of more risky cash crops.
4. Substantial volatility remains despite prospective risk management instruments considered. Farming is in general a risky business since crop yields and prices are relatively volatile in comparison to the expected farm income. In conjunction with a strike level set at for example 80% of the expected outcome, implying a deductible of 20%, explain the riskiness.
5. The budgetary implications of prospective risk management tools differed between case farms and future policy scenarios under consideration. To address the implications for budgetary expenses the payments involved with decoupled subsidy and insurance were studied. However, the budgetary consequences stemming from insurance depend on the level of granted premium subsidy. Total payments per hectare were affected by the alternatively policy scenarios. Gross premium payments on general field cropping farming systems, with more volatile cash crops, exceeded those on specialised cereals, oilseed and protein farms. Expected gross premiums under more liberal policies were more substantial than under protectionist policies offsetting the decoupled payments.

Concerning insurance decisions to cope with risks, the impact of possible factors influencing insurance purchase need to be discussed. These factors are farm and farmer personal characteristics (Mishra and Goodwin, 2003; Ogurtsov *et al.*, 2007; Sherrick *et al.*, 2004). The impact of farm characteristics as well as risk attitude were studied by means of case-specific analysis. However, risk perception was not yet addressed which is often regarded as a key farmer-specific factor – besides to risk attitude – to explain and model insurance purchase. Risk perception is defined as the mental interpretation of risk, decomposed as the chance to be exposed to the content and the magnitude of the risk. Decision-makers who perceived that a risk will relatively seldom occur will be less inclined to insure and self-insurance will be preferable. The participation decision is also negatively and significantly associated with the producer's belief about the availability of disaster relief in the future. So, if governments (continue to) provide free *ad hoc* disaster relief, an important incentive to participate will be severely undermined (Van Asseldonk *et al.*, 2002).

In general, a substantial demand for a market-based insurance product is not very likely conditional upon the relative high level of risk loading associated with insurance, moderate risk aversion and assuming that farmers don't perceive the perils as much more riskier than objective data would presume. Most of the adverse production years leading to potential indemnity payments under crop insurance cause only liquidity problems. Many arable farmers may well be able to 'ride out' the bad times by using savings or credit. Their access to such credit is likely to be good because they usually have substantial equity, mostly in the form of their investment in land (at least in North western EU member states). Use of insurance is likely to be of interest to such farmers only for catastrophic events which threaten the continuity of the firm, not for adverse years causing 'normal' income variation. Subsidising insurance schemes will increase potential participation. Subsidies can be provided for the farmer-paid premiums, for delivery and administration, and for the private sector reinsurance. It is obvious that many farmers will find it attractive to purchase crop insurance when the expected indemnities available exceed the cost of insuring. Serious questions, however, have been raised about their incentives (Skees, 1999).

# Chapter 7

# Policy options for risk management

*Ernst Berg and Joern Kramer*

## 7.1 Objectives

The objective of this chapter is to synthesise the work and findings of the Income Stabilisation project in order to come up with a set of policy options for risk management in the European Union. To achieve this, this chapter builds upon the results of all previous chapters as well as on the relevant literature and on recent work commissioned by EU institutions. In the first section, we give a brief review of types of risk and their main characteristics. This is followed by an outline of the risk exposure and risk perception of farms as it appears from the results of the respective chapters. The main part of the chapter deals with the evaluation of various risk management instruments, including theoretical aspects as well as empirical findings. This provides the basis for the proposal of policy options which are presented in the final section of the chapter.

## 7.2 Types of risk

Risk in general refers to the uncertainty of outcomes (Robison and Barry, 1987; cf Hardaker, 2000). It arises from the fact that the final result of the enterprise significantly depends on factors that are outside decision makers' control and cannot be predicted with certainty. The most important risks can be classified as follows (Hardaker *et al.*, 1997; Berg, 2005):

- *Production (or yield) risks*: the uncertainty of yields caused by the weather pattern of a specific year as well as by the emergence of pests or plant and animal diseases.
- *Price risks*: describe the fact that input or output prices exhibit unexpected changes after production or investment decisions have been made.
- *Asset risks*: include fire, theft and other damages or losses to farm assets. Most of these can be taken care of by commercial insurance.

- *Financial risks*: unpredictable changes of interest rates; significant liquidity fluctuations at high debt ratios.
- *Institutional (or policy) risks*: unexpected changes of laws and statutes (e.g. subsidies, tax regulations, environmental standards) that could have negative impacts on farm income.
- *Personal risks*: severe illness, injuries due to accidents or death of farm the operator or family members that work at the farm. Although generally insurable, these risks are often underestimated and receive insufficient attention, particularly in family farms.

These risks are often interrelated and all categories affect the outcome of business activities. Particularly outcomes at the lower part of the distribution may put the farm as a whole at risk if the effect is severe enough to preclude compensation through prior savings or borrowing.

Following Cafiero *et al.* (2005), in order to focus on the relevance of risk and the most appropriate risk management instruments to deal with it, the following distinction of risk generating events according to a three dimensional classification appears useful:
- *Frequency* of the event; from rare to frequent.
- *Intensity* of the impacts; from negligible to significant.
- *Correlation* between affected units; from idiosyncratic to systemic.

Cafiero *et al.* have put these three dimensions together in what they call a 'risk box' where each vertex corresponds to an archetypal risk (Figure 7.1).

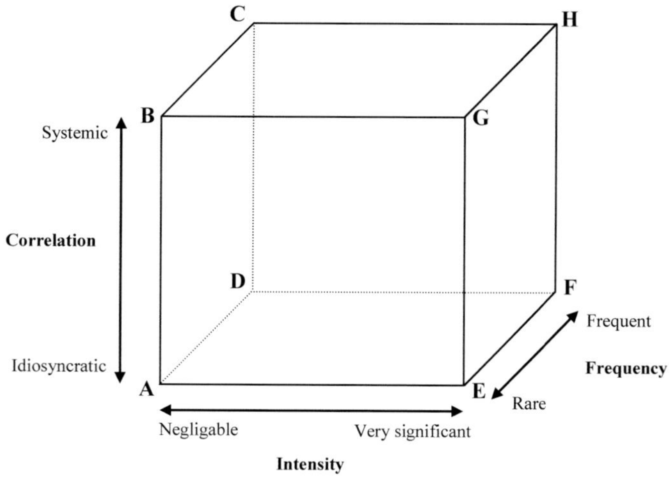


**Figure 7.1.** *Risk Box according to Cafiero* et al. *(2005, p. 4).*

While risks that have negligible impacts (vertexes A to D) in most cases could simply be retained, the other types require special attention. Risks located in the area of vertexes G to H, i.e. high impact, mostly correlated risks, are those which are likely to cause crises at farm level or even at sector level. We shall come back to the dimensions of the risk box during the discussion of risk management instruments at a later point.

## 7.3 Risk exposure and risk perception

### *7.3.1 Risk exposure of farms*

The final impact of risk is that it jeopardises the ability of households to smooth consumption over time. Risk exposure is therefore strongly related to income as the major source to satisfy consumption requirements, specifically to (unexpected) income fluctuations. The volatility of farm incomes was analysed in Chapter 2 based on FADN data. The analysis covers normal year to year income fluctuations and their main sources as well as income crises where the latter are illustrated referring to particular cases (swine fever, drought, BSE).

If the coefficient of variation (CV) is used as risk measure the farm level income volatility ranges from less than 0.3 to more than 0.6 depending on type of production and region. These figures suggest that there is potential for improvements in risk management. The financial robustness of farms in case of the occurrence of hazardous external events was assessed by analysing the ability to sustain a positive farm income after a 30% drop of total output. The figures show that in some regions only a small portion of specialised farms would meet this criterion indicating a high vulnerability to external shocks.

Overall, the conclusions from the analyses in Chapter 2 can be summarised as follows:

- The analysis of individual farm data shows strong fluctuations in farm income caused by yield variations due to climatic conditions and by the volatility of input and output prices.
- Furthermore, there are considerable differences between farm types and regions (member states). In most member states the intensive livestock sector exhibits the highest volatility of income which, according to the FADN data, is generally higher in Northern Europe than in the South.
- Farms in Northern Europe are more prone to hazardous external events, e.g. outbreak of contagious diseases or severe droughts, due to the structure of farming.

- The FADN is a useful data source to monitor farm income; however, its value as a tool to assess the need for (crisis) risk management actions by the government is limited. Main reasons for this include the lack of information on non farm resources and income, missing indicators for hazardous external events and the two years time lag before FADN data are available for all member states of the EU.

The main findings and conclusions with respect to the risk exposure of farms are also supported by other sources (e.g. European Commission, 2006a), and by the simulation results of Chapter 3.

### *7.3.2 Risk perception of farmers*

The demand for, and the use of risk management instruments after all depends on the farmers' subjective perception of the risk they feel to be exposed to, and on the benefits they belief to gain from the use of the respective instruments. Knowledge about the farmers' true perception of risk is therefore crucial when it comes to the design and evaluation of policy options. Chapter 5 aimed at providing this information through the results of a questionnaire. The survey was carried out in five EU member states and covered the areas of risk perception and risk experience as well as risk management strategies presently in use and planned for the future.

Summarising the results of the survey, one can state that the subjective perception of risks is largely consistent with the results from the data analysis (Chapter 2) and with theoretical deductions: Climatic effects and natural disasters are considered as the major sources of risk, followed by the volatility of prices. Differences in the relative importance occurred between member states and between different farm types, mainly according to the personal experience of the respondents. The farmers are also aware of farm management instruments at hand. These include risk sharing through various insurances and marketing contracts as well as on farm measures like diversification of production and income sources or holding financial reserves. Preventive actions like plant protection and technological improvements are perceived as particular effective instruments for risk reduction.

Taking into account the results of Chapter 5 and referring to other surveys documented in the literature (Meuwissen *et al.*, 2001; Musser and Patrick, 2002; cf. Akcaoz and Ozkan, 2005) one can safely state at this point that farmers are well aware of a number of risks they are exposed to. Among those that attract serious concern are weather effects, natural disasters and price risks, but also risks emerging from agricultural and environmental policy. It is perhaps worth noting that farmers in several countries (including the USA, Turkey and the Netherlands) rank the risks associated with changes in agricultural and environmental policies particularly high.

## 7.4 Risk management instruments

### 7.4.1 Classification of risk management instruments

Farmers have a wide variety of possibilities to influence the risk associated with their operations. Following Hardaker *et al.* (1997) and Berg (2005), these can be broadly classified into *on farm risk management instruments* on one hand and *market based or risk sharing instruments* on the other hand (Figure 7.2). The former include all measures that aim at avoiding or reducing the exposure to risks, such as precautionary actions to prevent accidents, fire outbreaks or burglaries. Furthermore, irrigation of crops or strategies to control pests and diseases in plant and animal production belong to this category. Spreading the risk through the diversification of farming activities is based on the fact that the dispersion of the overall return can be reduced by selecting a portfolio of activities that have outcomes with low or negative correlations. Finally, building financial reserves (including borrowing capacities) aims at creating a risk bearing potential that allows compensating the effects of unfavourable events, if necessary.

Referring to the risk box of Figure 7.1, it can be stated that strategies to prevent unfavourable events or at least reduce the possibility of their occurrence are relevant for all types of risk, provided that respective measures are available at affordable costs. Diversification strategies are useful for largely idiosyncratic or negatively correlated risks (i.e. the bottom side of the cube). They are particularly

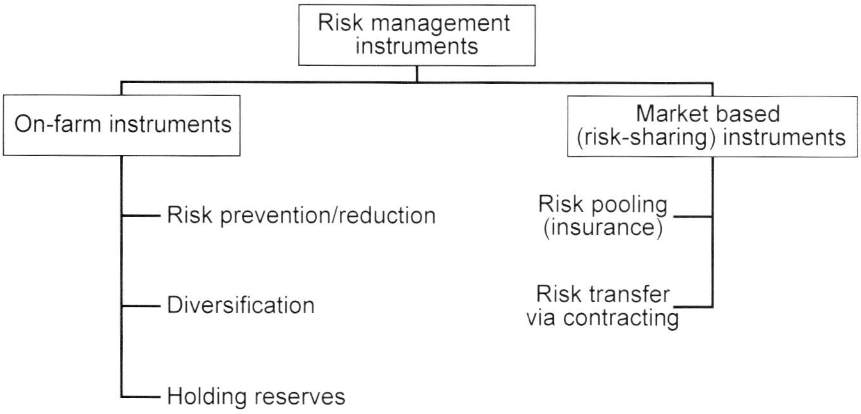


**Figure 7.2.** *Risk management instruments.*

beneficial if the uncertain events occur at high frequencies and have significant impacts (vertexes E, F). Relying on financial reserves is an option if the impacts of uncertain events on the overall performance are fairly limited. This, however, is probably true for most of what is often attributed as 'normal enterprise risk'. As Hardaker and Lien (2005, p. 12) state: '... at least in more developed countries where access to credit is easy, most farmers with reasonable equity can readily ride out normal year to year variations in income flows'.

Risk sharing instruments presuppose the existence of market partners. If risk pooling is possible insurance contracts may be the appropriate risk-sharing devices. In addition, risks can be shared with market partners by entering a contractual agreement. Popular examples include forward contracting of inputs and outputs as well as hedging with futures and options. Weather derivatives, often referred to as index insurance, also belong to this group.

The most popular risk sharing instruments are insurance contracts. However, insurance is a viable tool only if certain conditions are met as worked out in the large body of existing literature on this subject (see e.g. Ray, 1981; Hueth and Furtan, 1994; Rejda, 1995). The general conclusion from this is that insurance is a valuable instrument to manage such risks that are idiosyncratic, rare and cause significant damages (i.e. areas towards vertex E in Figure 7.1). Contrary, if damages are fairly low but occur frequently (vertex C) loss adjustment costs will largely reduce the attractiveness of insurance.

If risks are positively correlated (i.e. systemic) the pooling principle does not apply since most of the insured might claim indemnities at the same time, thus impairing the sustainability of the fund, unless the portfolio of risks is reinsured within a wider pool. For many production risks correlation is not a serious problem, given today's structure of the insurance industry and the possibilities to acquire reinsurance coverage. However, there are some exceptions: (1) contagious animal diseases and (2) extended droughts or floods in jeopardised regions, where the systemic nature of the risks may preclude commercial insurance solutions or make them unattractive because of high premium loading factors.

Examples for perfectly correlated risks are commodity prices since at efficient markets the individual producers' prices are all the same for a given commodity. Therefore the risk of unfavourable price changes will not be spread among the producers of that commodity (i.e. the top side of the risk box of Figure 7.1). However, since falling or rising commodity prices may affect different actors on the market in different directions, this provides the opportunity for the producers to share the risk by entering contractual arrangements with their marketing counterparts or by using financial derivatives like futures and options. For highly correlated production risks, options on regional yields (i.e. area yield insurance) or on weather indexes (i.e. weather derivatives) which belong to the same category of instruments are currently developing and have the potential to broaden the scope for the management of agricultural risks significantly.

All the above instruments induce costs either as direct costs, e.g. investment and operating costs of irrigation equipment, or as opportunity costs like the waiver of specialisation gains in case of a diversified production programme. Furthermore, they are interdependent in the sense that the effect of a certain measure on the overall risk exposure depends on the constellation of all other instruments. For instance, a broadly diversified production programme limits the benefit of additional risk management instruments. In principal, this requires an integrated approach to risk management which considers the full set of risk management instruments simultaneously to ultimately arrive at an optimal mix of instruments.

### *7.4.2 Evaluation of risk management strategies*

Following, we analyse the impacts that various risk management tools have on the risk exposure of farms. At the beginning of each section we address some theoretical aspects which are then followed by existing empirical results. Before dealing with the risk management tools themselves, a brief outline of important concepts of decision theory shall provide the methodological basis for evaluation of the instruments

#### Theoretical background

The most general approach for assessing the impacts of risky choices on a decision maker's well being is by means of expected utility. This requires that all possible outcomes of the risky prospect be translated into utility measures to compute the expected utility. Faced with a choice amongst a set of risky prospects, the expected utility hypothesis states that the prospect with the highest utility is preferred. The expected utility (EU) can be retranslated into a monetary measure, i.e. the certainty equivalent (CE), through the inverse of the utility function. The certainty equivalent represents the certain amount of money, which a decision maker with a given utility function would rate as equivalent to the uncertain outcome of the risky prospect (cf. Robison and Barry, 1987, p. 23ff). Ranking prospects by CE is equivalent to ranking them by expected utility.

By definition the certainty equivalent $CE$ equals the expected return $\mathrm{E}(y)$ minus the risk premium $\pi$, i.e. $CE = \mathrm{E}(y) - \pi$. For the latter Pratt has derived the approximate relationship $\pi = ½ R[E(y)] Var(y)$, where R[E($y$)] indicates the decision maker's absolute risk aversion measured at the expected value E($y$) and Var($y$) denotes the variance (cf. Robison and Barry, 1987, p. 34). Thus the certainty equivalent can be expressed as

$$CE = \mathrm{E}(y) - \frac{1}{2}\mathrm{R}\left[\mathrm{E}(y)\right]\mathrm{Var}(y) \tag{1}$$

yielding the well known expected value-variance (EV) approach which is often employed in portfolio analysis. The conditions under which the EV approach yields results consistent with the more general expected utility (EU) model have been worked out by several authors (cf. Meyer, 1987; Robison and Barry, 1987; Robison and Hanson, 1997). While Arrow (1971) originally argued that this was only the case under the premise of either quadratic utility or normally distributed income, Meyer (1987) has shown that EV orderings are consistent with the EU model for all distributions fully distinguished by location and scale, i.e. virtually all two-parameter distributions including log-normal, beta and gamma distributions. Besides its fairly general applicability, the EV approach has proven its usefulness because of its deductive strength. Many theoretical analyses are therefore based on this approach, including those we mainly refer to in this section.

In the following we specifically address aspects of diversification, standard and index-based insurance and hedging of price risks.

## Diversification as risk management instrument

Farms in Europe are typically set up as multi-commodity operations. While this may partly be based on agronomic grounds, it likewise has important impacts on the risk exposure as Berg and Schmitz (2007) have shown using the EV framework mentioned above. Referring to formula (1) the certainty equivalent is increased by the expected income and decreased by its variance. For simplicity let us assume that the expected returns of all activities are the same, so we can limit the analysis to inspecting the variance. Considering $n$ production activities realised in quantities $q_i$, the variance of income becomes

$$\mathrm{Var}(y) = \sum_{i=1}^{n} \sigma_i^2 q_i^2 + 2\sum_{i=1}^{n-1}\sum_{j=i+1}^{n} q_i q_j \,\mathrm{cov}_{ij} \tag{2}$$

where $\sigma_i^2$ represents the variance of the return of the $i$-th activity and $cov_{ij}$ denotes the covariance of returns between the activities $i$ and $j$. If we assume a portfolio of activities in which all quantities are equal, i.e. $q_i = 1/n$, the above equation becomes

$$\mathrm{Var}(y) = \frac{1}{n^2}\sum_{i=1}^{n}\sigma_i^2 + \frac{2}{n^2}\sum_{i=1}^{n-1}\sum_{j=i+1}^{n}\mathrm{cov}_{ij} \tag{3}$$

We now observe that a portfolio of n elements is comprised of n(n-1)/2 covariances. Thus, we can define an average covariance as

$$\overline{COV} = \frac{2\sum_{i=1}^{n-1}\sum_{j=i+1}^{n} cov_{ij}}{n(n-1)} \tag{4}$$

On substituting the second term of Equation 3 by this relation, the variance of the portfolio becomes

$$\mathrm{Var}(y) = \frac{1}{n^2}\sum_{i=1}^{n}\sigma_i^2 + \frac{n-1}{n}\overline{COV} \tag{5}$$

On introducing the average variance $\overline{\sigma}^2$ this equation further reduces to

$$\mathrm{Var}(y) = \frac{1}{n}\overline{\sigma}^2 + \frac{n-1}{n}\overline{COV} \tag{6}$$

Let us assume identically distributed returns for all activities. From $cov_{ij}=\sigma_i \cdot \sigma_j \cdot \rho_{ij}$ where $\rho_{ij}$ marks the correlation coefficient we can rewrite the average covariance as

$$\overline{COV} = \overline{\sigma}^2\overline{\rho}$$

where $\overline{\rho}$ marks the average correlation coefficient. Equation (6) then becomes

$$\mathrm{Var}(y) = \frac{1}{n}\overline{\sigma}^2 + \frac{n-1}{n}\overline{\sigma}^2\,\overline{\rho} = \frac{\overline{\sigma}^2}{n}\left(1+(n-1)\,\overline{\rho}\right) \tag{7}$$

The above equations indicate that the portfolio risk decreases as $n$ increases, however at diminishing rates (Figure 7.3). As the term $(n\text{-}1)/n$ approaches 1 for large $n$, the portfolio variance reduces to the average covariance which is not diversifiable. If the returns are stochastically independent, i.e. the correlation coefficients and covariances are zero, the risk is completely diversifiable. If the correlation coefficients amount to +1, no diversification effect occurs as can be seen from Equation 7. In turn, at correlation coefficients of -1 the portfolio variance completely vanishes already at $n = 2$.

For most cases the net returns from agricultural production activities are positively correlated, mainly because of the correlation of commodity prices. However, since these correlations are only moderate, significant diversification effects occur.

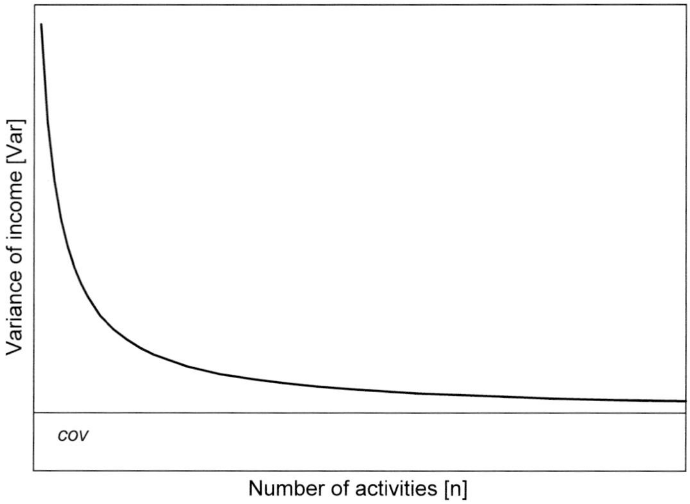


**Figure 7.3.** *Effect of diversification on the variance of income.*

For a farm with a broadly diversified production programme an additional risk management tool which further diversifies the portfolio is therefore of less value than for a highly specialised operation. In turn, if additional risk management tools become available they provide the opportunity to take more risk in other areas, possibly resulting in a higher degree of specialisation.

The model results from Chapter 3 and Chapter 6 largely reflect the theoretical derivations of the previous section. They clearly identify the diversification of the production programme as an important instrument to reduce risk. This statement is also supported by the results of Chapter 5 and other studies, e.g. Berg (2003); Berg and Starp (2006); Schmitz (2007); Mußhoff and Hirschauer (2008); Starp (2006). As a consequence, the impacts of all other risk management tools must always be assessed in the context of diversification.

### Insurance as risk management instrument

The basic idea of insurance is that of risk pooling, i.e. the insured faces the single risk of an unfavourable event with hazardous financial consequences while the insurer holds a large number of contracts which indemnify independent risks, thus expecting more or less the same indemnity payment every year because of the law of large numbers. If the insurer would charge the fair premium, i.e. the expected loss, the insured's expected income would remain unchanged while the risk would be transferred to the insurer. Assuming that risk aversion applies, the insured's willingness to pay exceeds the fair premium. Since the insurer has to cover overhead and transaction costs in addition to the indemnity payments, this is the necessary condition for an insurance market to develop. As the insured's willingness to pay depends on the reduction of risk and transaction costs are significantly influenced by the cost of loss adjustment, insurance solutions are particularly useful for risks that are characterised by low frequency and high impact.

To illustrate the potential demand for, and the impact of, insurance (particularly crop insurance) we use the state contingent representation of decisions under uncertainty according to Chambers and Quiggin (2000; 2004). In this approach, the uncertain production conditions are described as a set of states from which nature picks one independently from the decisions made prior to the disclosure of the true state of nature. If we presume only two states $z_1$ and $z_2$ for simplification, the approach can be illustrated graphically (Figure 7.4). Referring to Quiggin and Chambers (2004) the axes which represent the two states of nature are labelled 'drought' ($z_1$) and 'rain' ($z_2$). If a crop is produced without irrigation, yield will be high in the 'rain' state and low in the 'drought' state. Now assume that the crop is produced using an exogenously fixed bundle of inputs and generates a state-contingent income. One could then reallocate a portion of this income to provide irrigation facilities for a part of the acreage. This would yield an increased net income in the 'drought' state, however at the expense of sacrificing some income in the 'rain' state. On expanding the irrigated area, the state contingent net income would thus follow a transformation curve as depicted by the production/income frontier in Figure 7.4.

Now assume that the individual's objective function is to maximise expected utility $U$, given by

$$U = \pi_1\, u(y_1) + \pi_2\, u(y_2), \text{ where } \pi_1 + \pi_2 = 1 \tag{8}$$

In the above equation $\pi_1$ and $\pi_2$ denote the state probabilities. In case of risk aversion, for a given utility level $U$ the above function can be represented by an indifference curve as depicted in Figure 7.4. The slope of the indifference curve can be derived from the expected utility equation (Hirshleifer and Riley 1992, p. 44) and is given by

$$-\left.\frac{dy_2}{dy_1}\right|_{U=constant} \equiv \frac{\pi_1\, u'(y_1)}{\pi_2\, u'(y_2)} \tag{9}$$

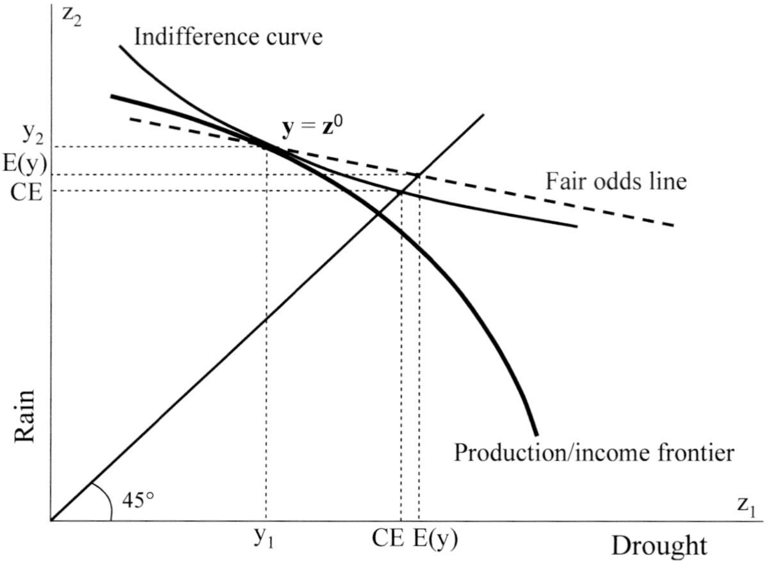


**Figure 7.4.** *State-contingent income and optimal production.*

where $u'(\cdot)$ denotes the first derivative of the utility function. The optimal choice is then given by the point of tangency between the individual's indifference curve and the income possibility set, i.e. at $y=z^0$. At this point the marginal rate of substitution between the state-contingent incomes equals the ratio of state probabilities weighted with the corresponding marginal utilities.

In Figure 7.4 the 45°-line through the origin marks all combinations of inputs that lead to the same net income in either state, thus representing a situation under certainty. The intersection of the indifference curve with the 45°-line therefore represents the certainty equivalent CE. Note that the absolute slope of the indifference curve at this point equals $\pi_1/\pi_2$ which is likewise the slope of the fair odds line that contains all combinations of state contingent incomes leading to the same expected income[8]. The latter is given by

$$E(y) = \pi_1 y_1 + \pi_2 y_2 \tag{10}$$

In Figure 7.4 the expected income $E(y)$ is graphically represented by the intersection of the fair odds line with the diagonal. Thus, the difference $E(y)$ – CE denotes the risk premium. Now consider the possibility to buy an insurance which pays an indemnity in the drought state ($z_1$) charging an insurance premium P. The loss

[8] As the income is he the same in either state, this holds true also for the marginal utilities $u'(\cdot)$ in Equation 9 which therefore cancel out.

associated with the occurrence of $z_1$ equals the difference $y_2 - y_1$. Assuming that the insured may choose a coverage level α, the indemnity payment in state $z_1$ amounts to $\alpha\,(y_2 - y_1)$. The expected income with insurance is then given by

$$E(y) = \pi_1\,[y_1 + \alpha(y_2 - y_1)] + \pi_2\,y_2 - P \qquad (11)$$

which in the case of full coverage (i.e. $\alpha = 1$) leads to a certain income of $y_2 - P$, corresponding to a point on the 45°-certainty line in Figure 7.4. Equation 11 represents the so called insurance line that can be further specified by determining how the premium is computed. Let us first consider a *fair premium*, i.e. one that equals the expected indemnity. In this case the premium amounts to $P = \pi_1\,\alpha\,(y_2 - y_1)$. On substituting $P$ in Equation 11 by this expression, the equation reduces to Equation 10, i.e. in the case of a fair premium the insurance line equals the fair odds line. Increasing the coverage level of the insurance would therefore move the point $y=z^o$ in Figure 7.4 on the fair odds line towards the diagonal, leading to an increasing expected utility. Any risk averse decision maker – regardless of the degree of risk aversion – would therefore acquire full coverage insurance to achieve the certain income E(y).

However, with the opportunity of transferring the risk through insurance, the decision maker's well being could be further improved by adjusting the production program. This effect is captured in Figure 7.5. Maximising the expected utility requires that the insurance line is tangent to both, the production/income frontier

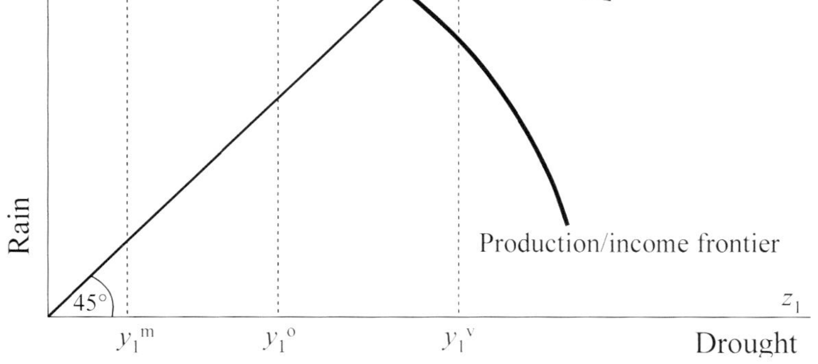


**Figure 7.5.** *Optimal production with insurance.*

and the indifference curve. In the case of a fair premium the absolute slope of the insurance line is $\pi_1/\pi_2$ leading to the production program represented by point $y^m$ on the transformation curve accompanied by full coverage insurance. The new production program maximises the expected market return and is more risky than the former one. In our example this means less irrigation yielding higher market returns in the 'rain' sate while the lower market returns in the 'drought' state are compensated by the insurance payout. Sine the fair premium equals the expected loss, the total net income amounts to the expected market return no matter which state of nature actually occurs.

Premiums charged by insurance companies are normally significantly above the fair premium. We therefore need to investigate the impact of premium loading. We first consider an additive surcharge. In this case the premium becomes $P = \pi_1 \alpha (y_2 - y_1) + c$, where $c$ represents the surcharge. Substituting into Equation 11 and rearranging the terms finally yields

$$E(y) = \pi_1 y_1 + \pi_2 y_2 - c \tag{12}$$

i.e. the insurance line is moved parallel towards the origin. As the tangency condition remains unchanged, full coverage insurance is still the preferred choice, however, the resulting (certain) income is lowered by the amount of the surcharge. Furthermore, the insurance remains attractive only as long as the surcharge does not exceed the risk premium. It therefore requires a certain degree of risk aversion to generate a demand for insurance.

Normally premium loading is expressed in relative terms, i.e. the total premium becomes $P = \pi_1 \alpha (y_2 - y_1)(1+\gamma)$, where $\gamma$ is the percentage surcharge so $(1+\gamma)$ represents a loading factor. Substituting this expression into (11) and rearranging the terms yields

$$E(y) = \pi_1 (1 + \alpha\gamma) y_1 + (\pi_2 - \alpha\gamma \pi_1) y_2$$

Regarding that $\pi_2 = 1 - \pi_1$ we obtain: (13)

$$E(y) = \pi_1 (1 + \alpha\gamma) y_1 + [1 - \pi_1 (1 + \alpha\gamma)] y_2$$

From this the absolute slope of the insurance line can be derived as

$$\frac{\pi_1(1+\alpha\gamma)}{1-\pi_1(1+\alpha\gamma)}$$

Compared to the fair odds line with the absolute slope $\pi_1 / \pi_2 \equiv \pi_1 / 1 - \pi_1$, the insurance line now exhibits as steeper slope. This means that, because of the tangency condition, partial insurance (i.e. $\alpha<1$) now becomes the optimal choice along with a production program that is somewhat less risky. In Figure 7.5 this would be represented by a point on the production/income frontier between $y^o$ and $y^m$.

It shall be noted at this point that in the above reflections we have always implicitly assumed that all conditions for the insurability of risks are perfectly met. These conditions refer to the problems arising from asymmetric information and to those associated with systemic risk. Although the theoretically derived conditions are never completely fulfilled in practice, it can safely be stated that the possibility to develop a commercial insurance market is greatly reduced if the distance from the ideal situation becomes too large.

Violations of the insurability conditions cause increasing transaction costs and thus lead to high premium loadings. As previously derived, significant premium loadings limit the attractiveness of insurance solutions to highly risk averse decision makers, and are therefore impediments for the development of insurance markets. Premium subsidies are *not* suitable to so solve this problem in the long run. One reason is that they – like all permanent subsidies – will be capitalised in the prices of production factors, e.g. land, and thus loose their intended effect in the course of time. Premium subsidies furthermore significantly affect the production programme as follows from the above analysis. They therefore distort markets and thus – at least from an analytical point of view – do not meet the 'green box' conditions of the WTO.

Agricultural insurance is offered within and outside of the EU in a variety of formats. Commercial insurance markets are fairly comprehensive in the fields of personal insurance, property insurance and liability insurance. In the field of production risks, besides some pure commercial insurance solutions for specific perils, e.g. hail, in many cases we find a high degree of public involvement (cf EC, 2006), particularly in the case of multiple perils crop insurance (MPCI). Often premiums are subsidised at high percentage levels. While many studies indicate that there are (potential and actual) benefits for the farmers, the evaluation of the overall effectiveness of highly subsidised MPCI systems is mostly negative; although significant improvements have occurred in recent years. The reasons for this include the systemic nature of the risks as well as the typical problems arising from asymmetric information, i.e. moral hazard and adverse selection, and last not last crowding out effects from other policy measures (Coble *et al.*, 2004; Skees, 2001). In cases where commercially viable premium obviously exceed the willingness to pay, it is questionable that insurance is the appropriate risk management instrument.

### Index insurance and weather derivatives

While standard insurance contracts confirm indemnity payments in case of the occurrence of a damage, index contracts base their payoffs on the value that an underlying index takes on. If the index is correlated with the revenue (or cost) of a firm, a contract that confirms a payment dependent on the value of the index can reduce income risk. Since the indexes are normally weather variables as rainfall or temperature, these instruments are also referred to as weather derivatives. They normally take on the form of option contracts.

Figure 7.6 depicts the payoff structure of a put option. The buyer of the option (long position) receives a payoff if the index $x$ falls below the strike level $K$. The payment then amounts to the difference ($K$-$x$), multiplied by a tick size that corresponds to the payment per index point. Deducting the premium $P$ from the payoff leads to the profit or loss of the option. The seller (short position) pays the claim and receives the premium. Thus, his profit is symmetrical to the long position.

This payoff structure corresponds to a Leontief type production function (cf Berg, 1997) that grows linearly with increasing $x$, until $x = K$, where the yield achieves its maximum. For the long position the payoff of the option is given by

$$A = V \cdot Max\left[\, 0, (K - x) \,\right] \tag{14}$$

where $V$ denotes the tick size. The fair premium $P_f$ of the option equals the discounted expected value of the payoff, E($A$), i.e.

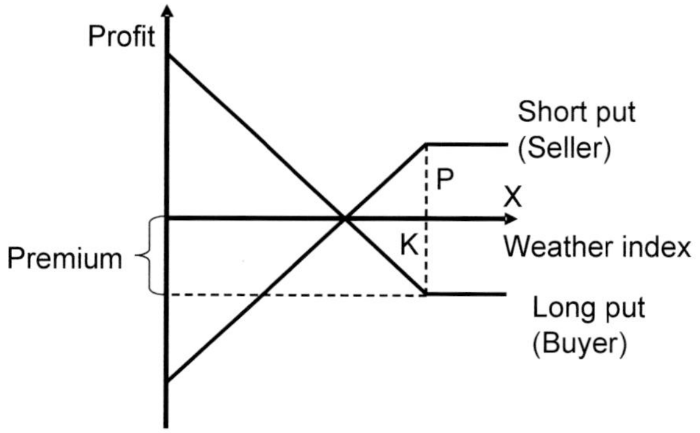


**Figure 7.6.** *Payoff structure of a put option.*

$$P_f = e^{-r \cdot d}\, \mathrm{E}(A) = e^{-r \cdot d}\, V\, \mathrm{E}(Max[\,0,(K-x)]) \tag{15}$$

where the factor $e^{-rd}$ discounts the payment over the duration $d$ using the interest rate $r$. The expected value of the *Max* function, E(*Max*[.]), represents the weighted average of the payments that occur if the index falls above or below the strike level $K$, respectively. Since no payment occurs at index values above $K$ we can write:

$$\mathrm{E}(Max[\,0,(K-x)]) = H(K) \cdot (K - \mathrm{E}(x \mid x \leq K)) \tag{16}$$

In Equation 15 $H$ marks the probability that $x$ falls below $K$. If $h(x)$ represents the density function of the weather index, then $H(K)$ is given by

$$H(K) = \int_{-\infty}^{K} h(x)\, dx \tag{17}$$

If the index is normally distributed, $H(K)$ becomes

$$H(K) = \Phi(z)\,, \quad \text{with} \quad z = \frac{K - \mathrm{E}(x)}{\sigma} \tag{18}$$

where $\Phi\,(z)$ represents the standard normal distribution. We still have to determine the expected value of $x$, given that $x$ falls below $K$ as represented by the term $\mathrm{E}(x \mid x \leq K)$. This is essentially the expected value of the distribution of $x$ truncated above $K$. The expected value of the truncated normal distribution is

$$\mathrm{E}(x \mid x < K) = \mathrm{E}(x) + \sigma \frac{-\phi(z)}{\Phi(z)} \tag{19}$$

where $\Phi(\cdot)$ is the standard normal distribution and $\phi(\cdot)$ the respective density function (Hartung, 1998, p. 149).

If the seller charges the fair premium, the expected revenue with or without the option remains the same for either position. The benefit for the buyer is the reduction of downside risk if the option payoff complements the market revenue. The precondition for this is a high correlation between the index and the production outcome. Following, we illustrate this using a numerical example.

The total net return per ha $W_p$ which comprises the market revenue plus the option payoff minus the fair premium $P_f$. is given by

$$W_P = y\, p_y + V \cdot Max[0, (\mathrm{E}(x) - x)] - P_f \tag{20}$$

In the above formula, $y$ is the yield and $p_y$ represents the product price. Now let $y$ represent the yield of wheat which we assume to be normally distributed with a mean of 80 dt/ha and a standard deviation of 10 dt/ha. If the product price $p_y$ is contractually fixed at 10 €/dt, with these assumptions, the distribution of the revenue corresponds to the solid black line in Figure 7.7 with an expected value of 800 €/ha and a standard deviation of 100 €/dt.

The weather index $x$ may represent the amount of rainfall during a certain period and shall likewise be normally distributed with a mean of $E(x) = 100$ mm and the standard deviation $s = 125$ mm. Setting the strike level at the expected value, i.e. $K = 100$, we derive the probability $H(100) = 0.5$ and the conditional expectation $E(x \mid x \leq K) = 90$ mm. Thus, the average negative deviation of the index from $K$, according to Equation 16, is 5 mm. Multiplying by a tick size $V = 8$ €/mm yields a fair premium of 40 €/ha[9]. Assuming that $y$ and $x$ are positively correlated random variables, the model of Equation 20 can be simulated stochastically. Figure 7.7 depicts the simulation results.

As can be seen from the graph, buying an option completely eliminates the downside risk, if and only if we assume a perfect correlation between the yield and the weather index. In this case the weather derivative is equivalent to an insurance contract based on the individual yield. In turn, at correlations less than +1 – even though they may be close to one – very low revenues cannot be excluded anymore, i.e.

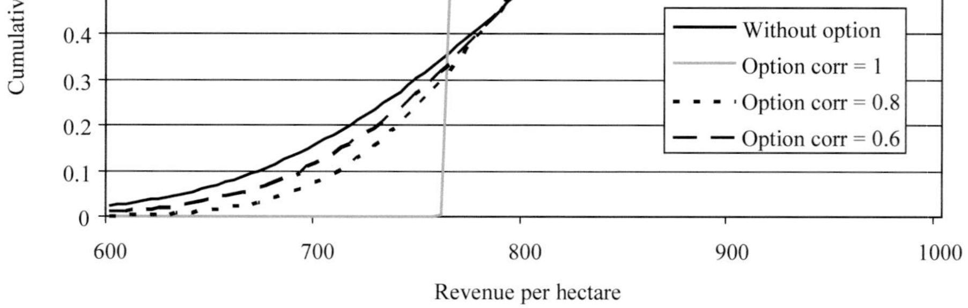


**Figure 7.7.** *Impacts of basis risk on the effectiveness of weather derivatives.*

[9] Since all payments are evaluated at harvesting time discounting is not necessary.

there is a remaining *basis risk*. Although the weather derivative always reduces the probability of low returns, it cannot secure a certain revenue because of the basis risk that is always present. This means that financial disasters caused by a local event, e.g. a hailstorm, flood or even pest damage, are still possible although fairly unlikely. Weather derivatives can therefore reduce profitability risks, but they cannot ensure liquidity. Likewise they cannot replace other types of disaster assistance.

Compared to individual crop or revenue insurance, the advantage of weather derivatives is that they are not prone to problems of moral hazard and adverse selection since the index cannot be influenced by the individual's behaviour. Furthermore, as standardised contracts can be traded at stock exchanges, risk can be shared between hedgers and speculators. This makes weather derivatives particularly useful for coping with systemic risks. Last but not least, low transaction costs (mainly caused by the absence of expensive individual claim adjustment procedures) make them valuable instruments for managing risks that have low impacts as single events, but occur at relatively high frequency.

Although still in the beginning, index-based derivatives are promising instruments. Recent examples include area yield insurance, index insurance and weather derivatives based on rainfall and other weather indexes. Up to now most applications relate to low income countries (Skees, 2007), however modelling studies indicate that there is also potential under European conditions (Berg and Schmitz, 2007; Torriani *et al.*, 2007; Mußhoff and Hirschauer, 2008). Thus there is likely to be a demand even in view of the remaining basis risk and with premium loadings comparable to standard agricultural insurance (cf. Mußhoff and Hirschauer, 2008).

### Hedging price risks

While weather derivatives are useful tools to hedge volumetric risks (input and output quantities), price risks can be hedged using futures and options. The basic mechanics imply that a commodity is sold for future delivery by means of a futures contract at its current price. The commodity itself is later sold on the cash market while the contract is bought back to lift the hedge. If the cash price and the future price develop equally until the delivery date, the risk of a price change between now and the delivery date will be completely eliminated through the hedge: if the price falls, the gain from the futures contract exactly offsets the decline of the cash price and vice versa. Since in reality cash and futures prices are not perfectly correlated – similar to the case of weather derivatives – a basis risk must be retained. Furthermore, hedging with futures involves some transaction cost. In total this means that the price risk cannot be completely eliminated and the risk reduction implies a cost represented by the decline of the expected revenue.

Let $y$ be the produced quantity, $f_0$ the current futures price (at time 0), and $p_1$ and $f_1$ the cash price and futures price at harvest (time 1), respectively. Then the net return π at time 1 is given by

$$\pi = p_1\ y + h\,(f_0 - f_1) - h\,\tau \tag{21}$$

where $h$ is the quantity of output hedged at time 0 ant τ marks the transaction costs. Thus the net return is comprised of the market revenue plus the gain from futures trading less the transaction costs. It is assumed that all production decisions were made prior to hedging, but the yield is still uncertain. When the farmer decides on $h$, he knows $f_0$ with certainty, but $f_1$ is correlated (imperfectly) with the uncertain cash price $p_1$, and so is uncertain.

Assuming that the joint probability distribution of the three stochastic variables is sufficiently characterised by their respective means, variances and covariances we can derive mean and variance of the net return. If cash and futures prices are not correlated with the yield (which is true for all relevant commodities except potatoes) the expected net return E(π) follows directly from Equation 21 on replacing the stochastic variables by their respective expected values E(·):

$$\mathrm{E}(\pi) = \mathrm{E}(p_1)\,\mathrm{E}(y) + h\,(f_0 - \mathrm{E}(f_1)) - h\,\tau \tag{22}$$

Assuming a deterministic yield for the moment, the variance of π is given by

$$\mathrm{Var}(\pi) = y^2\sigma_{p_1}^2 + h^2\sigma_{f_1}^2 - 2\,h\,y\,\rho_{p_1,f_1}\sigma_{p_1}\sigma_{f_1} \tag{23}$$

where $\sigma_{p_1}^2$ and $\sigma_{f_1}^2$ are the variances of $p_1$ and $f_1$, and $\rho_{p_1,f_1}$ is the correlation coefficient between cash price and futures price, so the last term of Equation 23 represents the covariance.

Introducing yield uncertainty makes the cash revenue a product of two stochastic variables. For a product of random variables, i.e. $z = x{\cdot}y$, the variance of $z$ can be computed according to the following formula (Bohrnstedt and Goldberger, 1969, p. 1439):

$$\mathrm{Var}(z) = \mathrm{E}(x)^2\sigma_y^2 + \mathrm{E}(y)^2\sigma_x^2 + 2\,\mathrm{E}(x)\,\mathrm{E}(y)\,cov_{x,y} + \sigma_x^2\sigma_y^2 + (cov_{x,y})^2$$

The above formula, in which $cov_{x,y}$ represents the covariance between $x$ and $y$, yields an exact measure of the variance if the probability density functions of the two random variables are symmetric. Otherwise the result is an approximation. If $x$ and $y$ are stochastically independent the formula reduces to

$$\mathrm{Var}(z) = \mathrm{E}(x)^2\sigma_y^2 + \mathrm{E}(y)^2\sigma_x^2 + \sigma_x^2\sigma_y^2$$

With this extension Equation 23 now becomes

$$\mathrm{Var}(\pi) = \mathrm{E}(y)^2\sigma_{p_1}^2 + \mathrm{E}(p_1)^2\sigma_y^2 + \sigma_y^2\sigma_{p_1}^2 + h^2\sigma_{f_1}^2 - 2\,h\,\mathrm{E}(y)\,\rho_{p_1,f_1}\sigma_{p_1}\sigma_{f_1} \tag{24}$$

Differentiating (22) and (24) with respect to $h$ yields

$$\frac{d}{dh}\mathrm{E}(\pi) = f_0 - \mathrm{E}(f_1) - \tau \tag{25a}$$

and

$$\frac{d}{dh}\mathrm{Var}(\pi) = 2\,(h\,\sigma_{f_1}^2 - \mathrm{E}(y)\,\rho_{p_1,f_1}\sigma_{p_1}\sigma_{f_1}) \tag{25b}$$

From Equation 25a we can see that – given the current futures price is an unbiased estimate of the futures price at harvest, i.e. $f_0 = \mathrm{E}(f_1)$ – hedging reduces the expected net return E(π) by the unit transaction cost. The variance Var(π) decreases as long as the term inside the brackets on the right hand side of Equation 25b is negative. The condition for this is

$$\frac{h}{\mathrm{E}(y)} \leq \rho_{p_1,f_1}\frac{\sigma_{p_1}}{\sigma_{f_1}} \tag{26}$$

In Equation 26 the relation $h/\mathrm{E}(y)$ represents the hedging ratio, i.e. the amount hedged as a portion of the expected yield. If the variances of cash and futures prices are of similar size, the hedging ratio up to which a risk reduction is possible corresponds largely to the correlation coefficient.

Based on Equation 1 and assuming constant absolute risk aversion (CARA), i.e. the utility function is $u(x) = 1 - e^{-\lambda x}$, the certainty equivalent of the net return can be expressed as

$$CE = \mathrm{E}(\pi) - \frac{\lambda}{2}\mathrm{Var}(\pi)$$

and from Equations 22 and 24

$$\begin{aligned} CE = {} & \mathrm{E}(p_1)\,\mathrm{E}(y) + h\,(f_0 - \mathrm{E}(f_1)) - h\,\tau \\ & - \frac{\lambda}{2}\left\{\mathrm{E}(y)^2\sigma_{p_1}^2 + \mathrm{E}(p_1)^2\sigma_y^2 + \sigma_y^2\sigma_{p_1}^2 + h^2\sigma_{f_1}^2 - 2\,h\,\mathrm{E}(y)\,\rho_{p_1,f_1}\sigma_{p_1}\sigma_{f_1}\right\} \end{aligned} \tag{27}$$

To find the optimal level of hedging, Equation 27 is differentiated with respect to $h$ and the result equated to zero:

$$\frac{dCE}{dh} = f_0 - \mathrm{E}(f_1) - \tau - \lambda\, h\, \sigma_{f_1}^2 + \lambda\, \mathrm{E}(y)\, \rho_{p_1,f_1} \sigma_{p_1} \sigma_{f_1} = 0 \qquad (28)$$

Solving for $h$ yields the optimal amount hedged as

$$h^* = \frac{f_0 - E(f_1) - \tau}{\lambda\, \sigma_{f_1}^2} + \frac{E(y)\, \rho_{p_1,f_1} \sigma_{p_1}}{\sigma_{f_1}} \qquad (29)$$

and the optimal hedging ratio as

$$\frac{h^*}{E(y)} = \frac{f_0 - E(f_1) - \tau}{\lambda\, E(y)\, \sigma_{f_1}^2} + \frac{\rho_{p_1,f_1} \sigma_{p_1}}{\sigma_{f_1}} \qquad (30)$$

The second term in Equations 29 and 30 represents the main hedging component: the lower the correlation between cash price and futures price (i.e. the higher the basis risk) the lower is the optimal hedging level. For a non producer (E(y) = 0), who can only participate in the market as a speculator, Equation 29 reduces to the first term. Thus, the first term can be regarded as speculative component (Pannell *et al.*, 2007). Speculation is encouraged (i.e. the first tern in Equation 29 is positive) if the expected settlement price $E(f_1)$ is lower than the current futures price $f_0$, and is discouraged by transactions costs. If the current futures price is an unbiased estimate of the settlement price, transactions costs lower the optimal hedging level, but this effect is moderated by high risk aversion.

As can be seen from the above formulas, increasing hedge ratios reduce the variance of revenue as long as the condition of Equation 26 is met. Without basis risk a 100% hedge ratio would lead to the minimum variance solution while increasing basis risk lowers the minimum variance hedge ratio. However, in the presence of transaction costs this risk reduction is only achieved at the expense of a progressive decline of the expected revenue.

Numerous studies in the USA have evaluated hedging strategies and illustrated their positive effects on risk reduction. Despite this, the use of this instrument by USA farmers is still fairly low. However, they make intensive use of cash forward contracting where the contracts are offered by the commodity vendors. These contracts, in turn, are only possible because the commodity traders themselves hedge their price risks on futures markets. In Europe, futures markets are still less efficient than in the USA because of low trading volume. Besides other factors, this is certainly due to the fact that the price support of the past decades made hedging with futures unattractive, especially in case of the big commodities. However, with ongoing liberalisation of markets this picture is likely to change in the future.

## 7.5 Policy options to support risk management

Public involvement in risk management – like in other policy areas – can only be justified through the intension to increase public welfare. The achievement of welfare gains implies that the benefits caused by the respective policy measures offset the associated cost, i.e. there is a potential for Pareto improvement.

Such welfare gains can generally emerge in two ways. One way refers to policy measurers to prevent the occurrence of crises – e.g. severe shortages of food supply in case of natural catastrophes – that would negatively affect large parts of the society. Besides this, reducing the volatility of individual incomes through public policy could be beneficial for the following reason: Under the assumption of widespread risk aversion, decision makers require a risk premium that increases with increasing volatility of incomes. Since this risk premium has the nature of transaction costs its reduction through income stabilising policies could increase welfare, given that the benefits from such policies offset the associated cost and that market distortions are negligible.

The above distinction likewise points to different types of risk, namely crisis risk on one hand and normal enterprise risk on the other, which is also the distinction Cafiero *et al.* (2005) make in their report. In the following, we adapt their notion and differentiate between policy options regarding the management of crisis risk and normal enterprise risk, respectively.

### *7.5.1 Policy options regarding crisis risk management*

Crisis risk generally refers to the consequences of a disastrous event. Adopting a United Nations definition, 'a disaster is a sudden, calamitous event that causes serious disruption of the functioning of a community or a society causing widespread human, material, economic and/or environmental losses which exceed the ability of the affected community or society to cope using its own level of resources' (cited in European Commission 2006, p. 24). In agriculture, such disasters can be caused by climatic events (severe droughts, floods, storms, etc.), pests (insects, snails, etc.) or diseases (foot and mouth disease, swine fever, etc.).

According to this general definition that largely corresponds to one given by the European Commission (cf. Cafiero *et al.*, 2005, p. 5), the conditions for qualifying a risk as being a crisis risk are

- that it happens suddenly, i.e. it is unforeseen,
- that it causes widespread losses, thus affecting a large number of units, and
- that it exceeds the individual capacity to cope.

In the risk box of Figure 7.1, crisis risks would therefore correspond to systemic and disastrous risks, represented by the G and H vertexes (Cafiero *et al.*, 2005, p. 5). From an individual point of view there is nothing that could be done to manage these risks except avoiding them by quitting the business in total. For public policy, there are basically two options:

- *Direct damage compensation* after the event has occurred is the only option in the short term. Following Cafiero *et al.* (2005), only damages to farm assets, such as buildings, equipment, green-houses, perennial crop stands, breeding livestock, etc., should be directly compensated.
- In the medium and long term there is also the possibility for *preventive actions* like public investments in protective infrastructure or the support of private actions that reduce the extent of damages caused by disastrous events. Presently, the world food system operates at historically low inventory levels and is therefore highly vulnerable to supply shortages due to bad weather conditions or other catastrophic events. In this situation, public investments in buffer stocks could be a valuable risk management policy. Preventive actions might also include measures that aim at establishing viable private markets for catastrophe insurance. *However, premium subsidies would not be included in this set because of the reasons mentioned in the earlier sections.*

More details are given by Cafiero *et al.* (2005, p. 46). An important point is that rules are set at EU level stating the conditions under which disaster relief according to (1) will be granted (i.e. type of event, extent of losses, and proportion of the loss that is compensated).

### *7.5.2 Policy options regarding normal enterprise risk*

In general, farmers do have a rather comprehensive set of instruments available to manage normal business risk. However, while on farm instruments (Figure 7.1) seem to be more or less fully developed, well known by farmers and farm advisors, and thus are widely used, this does not equally apply in the case of market based instruments. Commercial insurance markets are fairly comprehensive in the fields of personal insurance, property insurance and liability insurance, while various production risks remain uncovered. Deficiencies mainly relate to risks that are systemic or at least partially systemic in nature, e.g. yield uncertainty due to climatic events, contagious animal diseases, and price risks. Systemic risks can be effectively tackled using derivative markets. In Europe, probably due to the long lasting price support policy, commodity futures markets are still underdeveloped and hardly used by farmers. The same is true for other derivatives markets. As a consequence, there is only little knowledge and hardly any experience in the agricultural sector as to how these markets could be used to effectively hedge business risks. Therefore there is a need for public policy measures intending to facilitate the operation of private markets, including insurance as well as other financial instruments. More precisely such measures could aim at:

- Education of farmers and extension personnel in risk management issues, particularly in the functioning and the use of derivative markets.
- Support of the development of private insurance/derivative markets (e.g. index-based insurance or weather derivatives) without paying premium subsidies. Support may include
  - providing the regulatory institutions and informational support,
  - the development of informational infrastructure (monitoring equipment and databases),
  - direct participation in the market during the starting phase, e.g. by offering options based on weather indexes, or by providing public re-insurance,
  - other forms of start-up support.
- Support of the development and operation of mutual funds. Public policy could provide matching contributions to those of the farmers and set up the rules for funds' withdrawals. This can be a viable option to securitise production risks in the case of specialty crops or animal diseases.
- Support of institutional arrangements in the sense of public private partnerships that provide risk management services to the farmers. Since the selection of an optimal portfolio of risk management tools is a complex task it can be doubted that farmers – besides all other tasks they have to accomplish in their predominantly small to medium sized operations – will ever be able to cope with this problem. Instead, special institutions could take over the task of creating and managing such portfolios that fit the need of particular farm types. The farmers themselves would then only have to deal with one aggregate instrument aimed at reducing their downside risk of income.

# Chapter 8

# Epilogue

At the final project meeting in Warsaw, February 2008, two steering committee members were asked to reflect on the Income Stabilisation project. Their views are different with respect to whether there should be premium subsidies. Views converge with regard to the need for further analyses in the international agricultural risk management arena.

## 8.1 Reflections by Carlo Cafiero, Warsaw, February 2008

Reflections were structured along the lines of the various components of the Income Stabilisation project:

1. Risk exposure:
   - There is a need for household level income data; FADN is not sufficient.
   - Indications on the cost of risk exposure could be obtained indirectly through *expenditure* patterns of farm households and bankruptcy rates.
   - What about tax systems? Are there tax returns by farmers? Do they 'match' FADN data?
   - Academicians should suggest ways of improving FADN.
   - With regard to future scenarios we should anticipate on patterns of correlation (which is difficult), structural adjustments and behaviour intended to reduce risk exposure.
2. Risk management experience and perception:
   - Is there any country where multiple peril crop insurance is available without public subsidy?
   - Does the subsidy to the premium solve *any* of the market failures that plague insurance?
   - If governments want to promote insurance, they should address the real market failures, such as information asymmetry and monitoring issues.
   - 'Insurability' of the risk, that is the extent to which market-based insurance is capable of effectively spreading the cost related to a certain risk, should

be considered more carefully before suggesting that a subsidy may promote participation.

3. Economic impact and policy options:
   - With respect to identifying farm-level impact of risk management instruments, the whole-farm, i.e. portfolio approach, is well chosen. Even better would be a whole household, consumption smoothing approach.
   - How to deal with the asset value, i.e. the 'wealth effect'?
   - Policy options should consider the potential for index contracts.
   - Basis risk can be pooled as it is idiosyncratic by definition.
   - Virtuous behaviour, i.e. the incentive to protect your yield is still there – no moral hazard.
   - Innovative contracts should consider the difference between price and yield risk. A low price is a benefit to the purchaser of the product, therefore the aim is how to have them to 'chip in' the risk management pool.
4. Concluding remarks:
   - Why aiming at policy intervention for protecting farmers from risk exposure?
   - New issues that have not been covered: liability risk for product quality, environmental damages, contractual breaching risk, value chain disruptions.

## 8.2 Reflections by Nikiforos Georgiadis, Warsaw, February 2008

After 3 years of research, discussions and hard work the Income Stabilisation project is reaching its completion. Scientists from established and new member states worked together to collect information related to: (1) risks threatening farmers' income in a number of selected member states; and (2) policies and established methods of managing these risks and related tools used by the farmers both in EU and other parts of the world. Data were analysed and processed to quantify the risk exposure of farm households. Furthermore, the WTO and CAP impact on farmers' risk exposure and risk management opportunities were examined. A whole-farm model is developed which provides insight into the impact of risk management instruments on farm income volatility and farm crisis risk. Finally, a synthesis of the above works resulted in a list of viable risk management tools including recommendations for designing and implementation issues.

It is worthwhile to point out that a considerable amount of information needed for the works of the project was raised from previous studies such as the 'risks and crisis management in agriculture' requested by the European Parliament and prepared by Carlo Cafiero et al., or the 'Agricultural Insurance Scheme' requested by the European Commission and prepared by the Joint Research Centre in Ispra, Italy, as well from official statistical sources, such as the FADN records and by conducting original research about farmers perceptions of risk exposure and risk management instruments. (The latter is an interesting work that revealed some not anticipated perceptions but also verified some of our century old beliefs shaped in our university times, about farmers' attitudes).

I am sure that in the future the conclusions and the recommendations made by the project team will become a reference for policy makers and the methodology and analyses conducted will also become a reference for the academic people working in the field of risk management in agriculture.

I believe that much of the work done is applicable to the member states, though to a varying degree, as each country has developed its own risk management tools which may not be the same as those developed in other member states.

The analysis carried out by the project team and the work done by other researchers and authors in the past, confirm that there are considerable climatic, farming practices and farm types differences among regions and countries in the EU as well as among the on and off-farm risk management strategies followed by the EU farmers and the member states.

If one adds to this, the complexity of the risk profile of a farm, he/she easily arrives at the conclusion that flexibility in farm risk management policies at a European level is a must.

Policy flexibility, among other things implies the existence and availability of as many as possible, risk management tools. Indeed, the project team has worked out and presented a considerable number of instruments that could be used by farmers.

In the field of risk transfer or risk sharing, traditional insurance, index insurance and weather derivatives as well as hedging price risks with futures and options and cash forward contracting are discussed. Although the non traditional tools are praised by the project team for their effectiveness, it is admitted that their use is still fairly low in their birthplace, the USA. In the EU, only a couple of pilot index insurance programmes are run whereas weather derivatives are used by reinsurers (e.g. Swiss Re, Munich Re) in order to reinsure insurers offering traditional insurance to their clients. (That means, that the market for non-traditional risk transfer management tools used by farmers, virtually, does not exist in the EU).

Therefore, taken into account the limited use of these tools in the USA (much earlier introduced there), the existing efficiency gap between European and American futures markets, the small to medium European farm size and the diversity in the farm types recorded in the member states (which are characteristics that increase the basis risks and therefore make the above mentioned tools less attractive), one could argue that the market of non-traditional risk transfer or sharing tools in the short and medium term (say a period of at least 10 years) will remain a niche market.

This implies that the traditional insurance market will remain a mainstream market and the traditional insurance will remain one of the main financial tools for farmers' risk transfer during the next decade.

The above view is also supported by the finding, presented in the final chapter about farmers' perceptions (Szekely and Palinkas), that the majority of the farmers (more than 60%) 'would continue using the currently applied method in the future'.

Taking the above into consideration, it would be wise to strengthen the traditional insurance to cope with an increased need for risk transfer in the next, at least, decade during which, however, any effort has to be made in order to develop non-traditional tools.

In the policy options draft paper of this project, one can find recommendations for establishing viable private markets for catastrophe insurance or for improving the regulatory and supporting infrastructure or for direct intervention/participation in the market (developing non-traditional risk transfer tools or providing reinsurance). My feeling is that very few people would disagree with these recommendations, but I would like to point out that they are formulated in a rather qualitative form and therefore their adequacy and efficiency regarding their impact on insurance market development cannot be fully evaluated in the framework of this project.

However, it is explicitly stated, that in all cases, private insurance should be developed without paying premium subsidies. The key words for arriving at this absolute and firm position are: rent seeking, efficiency, tax payers, and a little of compliance. But one could argue that in our free market economies on both sides of the Atlantic Ocean, state intervention is not unknown at all. Various types of state subsidies given to various stakeholders make the free competition a little bit less free and more distorted. If we wanted to make decisions on a pure scientific basis, either we should abolish all of them (which, in practise looks impossible) or a holistic comparative study should be carried out in order to compare the various types of public spending, i.e. subsidies, so that the policy makers are in a position to decide which one is less efficient or more susceptive to rent seeking or constitute a greater burden on the shoulders of the tax payers. To my knowledge, there is not available sufficient work of this type and therefore one could argue that it is extremely risky to base the relative conclusions of this project on a limited existing literature on the subject. There is not available sufficient work even for those issues used to put the blame on the premium subsidies. (According to the findings of this research project we have – world wide – no evidence, but only indications 'that the insurance companies end up capturing the rents, i.e. profits, resulting from the premium subsidies' (Garrido and Bielza, 2008). Therefore, a thorough quantitative analysis about rent seeking, has to be carried out before making any decision, based on this argument, to exclude premium subsidies from risk management policies.)

I would be more prepared to discuss Carlo Cafiero's view that the EU, in order to subsidise by 50% the premium of a multi-peril European crop insurance programme, would have to pay such an amount of money that it may not be available under the prevailing conditions in the European Union.

Nevertheless, we do not need to examine the issue of premium subsidies looking at the extremes (nothing or everything). The project team is not doing this when proposing reinsurance as a possible risk management instrument. They would have rejected reinsurance on the same basis (of that of premium subsidies mentioned above) if they had estimated the capacity needed for a full reinsurance coverage both for the crop and livestock sector, similar, for instance, to that offered by the 'Consorcio' (CCS) in Spain. I believe they have nothing like this in mind. I believe they are thinking something smaller. That is, also, the way we have to follow when examining the premium subsidies issue. Think small. Think in a flexible way. Let's talk about premium subsidies in countries that just start developing crop and livestock insurance. Or let's talk about premium subsidies for the introduction of new agricultural insurance products, not available in the market currently, or newly introduced agricultural insurance schemes, or let's talk about much smaller than 50% EU participation in premium subsidies of established programmes (the remaining covered by the member states). Finally, let's talk about premium subsidies as an option among other options that are available within certain budgetary constraints. Then, the amount of money required will become much smaller and we will be in a position to examine from a realistic point of view the possibility to provide it.

Talking about premium subsidies, it would be worthwhile to remind the findings presented in the work by Garrido and Bielza, that: (1) the percentage of the insured crop production in the member states with no insurance subsidies accounts for only the one fourth of that in the member states with insurance subsidies; and (2) growing insurance portfolios increase the effectiveness of risk pooling and reduce the cost of reinsurance).

I would add, from a point of view of a practitioner that sizeable insurance portfolios allow for economies of scale, investments in know-how and improved quality of services.

Following the previous discussion I could argue that subsidised insurance should be a fundamental option in the EU risk management policy and it should be placed in the same order of importance with reinsurance. However, I would like to add a few more arguments for this, related to the financial sector and the EU-WTO new regulatory environment that will prevail in the next decade.

The regulatory environment within which the financial sector will be obliged to operate in the immediate future is becoming harder. Banks have to work within the Basel II framework. More strict rules apply now, for defining a delay in the repayment of a loan as a default. Increased numbers of defaults may pose a threat to the credit risk models of the banks. Because such a development may lead to increased regulatory capital requirements, the banks would be extremely disliked and would counteract to eliminate the possibility of experiencing a situation like this by providing credit under more strict rules to farmers that are not protected adequately by a safety net. Following this, more and more farmers will face the danger of a less accessible and more expensive credit market. A combination of a restricted and costly access to credit and of a not adequate safety net will lead to an increasing number of farmers being ruined. Premium subsidies will increase farmers' capability to get insured a fact that will improve their access to credit and will reduce the possibility of being declared defaulted and finally bankrupted after a disaster.

Solvency II, is a coming regulatory framework, within which the insurers will be forced to operate by the end of 2012 (at the latest). In this type of regulatory framework, insurance products of higher volatility may lead to greater regulatory capital requirements (MCR or SCR) and therefore they will be less attractive to insurers. Thus, an insurer will have an additional (to the existing) reason to consider as more attractive, for instance, car insurance than multiple peril crop insurance. Solvency II will contribute to more nervous and impatient markets for high volatility insurance products, such as crop insurance, and to fewer insurers writing agricultural insurance, a fact that may lead to higher premiums.

Premium subsidy can work as a stabilising factor in those markets and help in keeping inside the market both insurers (by assisting them to built a sizeable agricultural insurance portfolio within which they could more efficiently diversify the risk in space and in time and therefore to reduce their portfolio volatility) and farmers (by reducing their insurance cost, which in agricultural insurance is often unbearably high, and thus allowing them to stay insured at an affordable cost). It should be reminded that insured farmers have easier and less costly access to credit.

Availability of reinsurance will have a similar, but not equal, effect directly on the side of the insurers and indirectly on the side of the farmers.

Last but not least, the EU regulatory environment related to state aid provided on the basis of articles 87 & 88 of the Treaty is changing according to the Commission Regulation (EC) No 1857/ 2006 and the 'Community Guidelines for State Aid in Agricultural and Forestry Sector 2007 – 2013'. In article 11 (point 8) of the above regulation it is stated that 'From 1 January 2010, compensation offered must be reduced by 50% unless it is given to farmers who have taken out insurance covering at least 50% of the average annual production… and the statistically most frequent risks...'

In a number of member states agricultural insurance is underdeveloped and most of their farmers are a little acquainted or not acquainted at all with crop or livestock insurance and, in addition to this, they are discouraged to be insured due to high premium rates and low quality services. (A narrow insurance market is expensive. A narrow agricultural insurance market is unbearably expensive).

Many of the said farmers, not being insured according to the above regulation, will suffer losses for which they will not be eligible to receive *ad hoc* compensation greater than the 50% of what they would be paid before the 1/1/2010.

Taking into consideration that: (1) the percentage of the insured crop production in the member states with no insurance subsidies account for only one fourth of that in the member states with insurance subsidies; AND (2) according to our experience, the insurance markets need long time to develop and this time is much longer without subsidising the premium, one could argue that the above countries will have no chance to make available to their farmers, in the mid term, even single peril insurance products in some of the agricultural branches, without adopting policies that incorporate premium subsidies.

Without such policies the market may remain small, expensive and poor in products' quality and availability for many years, and less transparent – as insurers, operating independently, rarely publicise technical information (which is not the case in subsidised schemes) – even if this market is supported by other measures proposed by the project team. In such a case the farmers of those countries will be asked either to buy low quality, expensive products or to retain a big part of the catastrophe risks. In the first case a considerable number of them will fail to remain insured and in the second case an increasing number of farmers will become bankrupted after each occurrence of a natural disaster.

As a last note to this section of my speech, I would like to point out that the WTO regulatory framework regarding agricultural insurance may also change. To my knowledge the Americans, in the framework of the present round of the WTO negotiations, push towards a revision that may put the premium subsidies in the Green Box. If such a revision takes place more countries will incorporate premium subsidies in their agricultural risk management policies in an attempt to make their farms more competitive.

Recapitulating, I say that although I believe that premium subsidies present advantages that have been disregarded in this study, I would agree that they also do generate problems. However, because I have right now a terrible headache, I will not cut my head off. I would rather take an aspirin or another appropriate medicine. Thus, I maintain that the premium subsidies should be incorporated in the recommended policies (as an option), in a flexible way, so that insurance could play its significant role as a risk management tool in the next decade. Towards the same goal, the contribution of reinsurance will be decisive and its role should be emphasised in the final document that will reflect the brilliant work done in the framework of the Income Stabilisation project.

Thank you for your patience.

# References

Agra Europe Weekly: various issues Doha Round effect absent from upbeat EU market report (10 February 2006) Doha Round: Sensitive products clause essential to EU dairy sector (03 March 2006), EU Summit, WTO Ministerial point way ahead for CAP (23 December 2006).

Agresti, A., 2002. Categorical Data Analysis. 2nd edition. John Wiley & Sons, USA.

Akcaoz, H. and B. Ozkan, 2005. Determining risk sources and strategies among farmers of contrasting risk awareness: a case study for Cukurova Region of Turkey. Journal of Arid Environments, 62, 661-675.

Alizadeh, A. and N. Nomikos, 2005. Agricultural reforms and the use of market mechanisms for risk management. Futures and Options Association, London, UK.

Allanson, P. and L. Hubbard, 1999. On the comparative evaluation of agricultural income distributions in the European Union. European Review of Agricultural Economics, 26, 1-17.

Anderson, J.R., J.L. Dillon and J.B. Hardaker, 1977. Agricultural decision analysis. Iowa State University Press, Ames, USA.

Anton, J. and C. Giner, 2005. production incentives of risk-reducing policies and strategies. Organisation for Economic Co-operation and Development, Paris, France.

Arrow, K.J., 1971. Essays in the theory of risk bearing. Markham, Chicago, USA.

ATTRA /National Sustainable Agriculture Information Service, 2005. Risk management survey. Accessible at: http://www.attra.org/risk_management/Survey/rmsurveyeng.doc.

Barnett, B.J., J.R. Black, Y. Hu and J.R. Skees, 2005. Is area yield insurance competitive with farm yield insurance? Journal of Agricultural and Resource Economics, 30, 285-301.

Benjamin, C. and A. Kihmi, 2006. Farm work, off-farm work, and hired farm labour: estimating a discrete-choice model of French farm couples' labour decisions. European Review of Agricultural Economics, 33, 149-171.

Berg, E., 2003. Modelling the Impacts of Uncertainty and Attitudes towards Risk on Production Decisions in Arable Farming. In: '25th International Conference of IAAE, August 16-22, 2003'. Durban, South Africa. Online (http://www.lbwl.uni-bonn.de/pu/Forschung/Publikationen/IAAE2003_BE.pdf).

Berg, E., 2005. Integriertes Risikomanagement – Notwendigkeit und Konzepte für die Praxis. In: Deitmer, J. (ed.), 'Agrarökonomie im Wandel, Tagungsband anlässlich des 80. Geburtstages von Prof. em. Dr. Dr. h.c. Günther Steffen am 24. September 2004, 53-67'. ILB-Verlag, Bonn, Germany.

Berg, E. and B. Schmitz, 2007. Weather based instruments in the context of whole-farm risk management. In: 'Management of climate risks in agriculture, 101st EAAE Seminar, July 5-6, 2007'. Berlin, Germany.

Berg, E. and M. Starp, 2006. Farm level risk assessment using downside risk measures. In: '26th International Conference of the IAAE , August 12-18, 2006'. Gold Coast, Australia, Online at AgEcon Search (http://agecon.lib.umn.edu/).

Berkhout, P. and C. van Bruchem, 2006. Landbouw Economisch Bericht 2006. Report PR.06.01, LEI, The Hague, The Netherlands.

Bohrnstedt, G.W. and A.S. Goldberger, 1969. On the exact covariance of products of random variables. Journal of the American Statistical Association, 64, 1439-1442.

Burgaz Moreno, F. and M.M. Pérez-Morales, 1996. 1902-1992 Noventa Años de Seguros Agrarios en España. Serie Estudios, MAPA, Madrid, Spain.

Cafiero, C., F. Capitanio, A. Cioffi and A. Coppola, 2005. Risks and crisis management in agriculture. European Parliament document, IP/B/AGRI/ST/2005-30.

Chambers, R.G. and, J. Quiggin, 2000. Uncertainty, Production, Choice and Agency: The State-Contigent Approach. Cambridge University Press, New York, USA.

Chambers, R.G. and J. Quiggin, 2004. Technological and financial approaches to risk management in agriculture: an integrated approach. Australian Journal of Agricultural and Resource Economics, 48, 199-223.

Chambers, R.L. and C.J. Skinner (ed.), 2003. Analysis of survey data. John Wiley & Sons, UK.

Coble, K.H., J.C. Miller, M. Zuniga and R. Heifner, 2004. The joint effect of government crop insurance and loan programmes on the demand for futures hedging. European Review of Agricultural Economics, 31, 309-330.

Court of Auditors, 2002. Special Report No 14/2003 of the Court of Auditors concerning agricultural statistics measurement of farm incomes.

Dismukes, R., and R. Durst, 2006. Whole-farm approaches to a safety net. Economic Information Bulletin, 15.

EU DG Agri, 2007. Agriculture in the European Union – Statistical and economic information 2006. http://ec.europa.eu/agriculture/agrista/2006/table_en/index.htm.

European Commission, 2001. Risk management tools for EU agriculture, with a special focus on insurance. Working Document, Agriculture Directorate-General.

European Commission, 2005. Communication from the Commission to the Council on risk and crisis management in agriculture. COM 74. Brussels. Belgium.

European Commission, 2006a. Agricultural insurance schemes. JRC report. Ispra.

European Commission, 2006b. Evaluation of the Community Animal Health Policy (CAHP) 1995-2004 and alternatives for the future. Brussels, Belgium.

European Parliament, 2008. Response of the European Parliament on CAP health check. http://www.europa-nu.nl/9353000/1/j9vvh6nf08temv0/vht5nb9sauzs?tab=2&ctx=vg9pir5eze8o.

Eurostat, 2002. Income of the agricultural households sector 2001 report. Theme 5. Eurostat, Luxembourg.

EUROSTAT, 2003a. Structure and typology of agricultural holdings. Working document, Luxembourg.

EUROSTAT, 2003b. Farm structure survey 2003. http://epp.eurostat.ec.europa.eu.

EUROSTAT, 2005. F.A.D.N. http://ec.europa.eu/agriculture/rica/index_en.cfm.

FAO, 2004, The state of agricultural commodity markets. Rome, Italy.

FAPRI, 2005. U.S. Proposal for WTO agriculture negotiations: its impact on U.S. and world agriculture, CARD Working Paper 05-WP 417, December 2005.

Findeis, J., 2002. Subjective equilibrium theory of the farm household: theory revisited and new directions. In: 'Workshop on the farm household -firm unit: its importance in agriculture and implications for statistics'. Penn State, Imperial College, ERS. <http://household.airs.psu.edu>

Forteza del Rey, V., 2002. Diagnóstico del Seguro Agrario en los Diferentes Países de la Unión. In: 'International Conference: Agricultural Insurance and Income Guarantees, Madrid 13-15 May'. Madrid, Spain.

Freund, R.J., 1956. The introduction of risk into a programming model, Econometrica, 24, 253-263.

Garrido, A. and M. Bielza, 2008. Review of historic, current and developing private, public and public-private risk management instruments. Department of Agricultural Economics, Technical University of Madrid, Spain.

Garrido, A. and D. Zilberman, 2007. Revisiting the demand for agricultural insurance: the case of Spain. In: 'Management of climate risks in agriculture', 101st EAAE Seminar, July 5-6, 2007. Berlin, Germany.

Goodwin, B.K. and A.P. Ker, 1998. Revenue insurance – a new dimension in risk management. Choices, Fourth Quarter, 24-27.

Gundersen, C., M. Morehart, L. Whitener, L. Ghelfi, J. Johnson, K. Kassel, B. Kuhn, A. Ishra, S. Offnutt and L. Tiehen, 2000. A safety-net for farm households. Agricultural Economic Report, 788, ERS, USDA, Washington, DC.

Halcrow, H.G., 1949. Actuarial structures for crop insurance. Journal of Farm Economics, 31, 418-443.

Hardaker, J.B., 2000. Some issues in dealing with risk in agriculture. Working Paper Series in Agricultural and Resource Economics No. 2000-3. University of New England, Armidale, Australia.

Hardaker, J.B. and G. Lien, 2005. Towards some principles of good practice for decision analysis in agriculture. Working paper 2005-1. Norwegian Agricultural Economics Research Institute, Oslo, Norway.

Hardaker, J.B., R.B.M. Huirne and J.R. Anderson, 1997. Coping with Risk in Agriculture. CAB International, Oxfon, UK.

Hardaker, J.B., R.B.M. Huirne, J.R. Anderson and G. Lien, 2004. Coping with Risk in Agriculture. CAB International, Oxfon, UK.

Harmignie, O., P. Polomé, B. Henry De Frahan and F. Gaspart, 2005. Agricultural yield and price distributions. In: '8th ABER-BLVE Workshop, Ghent, 27 april, Association belg d'économie rurale ed'.

Harrison, A., 1975. Farmers and farm businesses in England, Misc. Studies 62, Department of Agricultural Economics and Management, University of Reading.

Hartung J., 1998. Statistik: Lehr- und Handbuch der angewandten Statistik. Oldebourg, München, Germany.

Hazell, P., C. Pomareda and A. Valdés, 1986. Crop insurance for agricultural development: issues and experience. Published for IFPRI by Johns Hopkins University Press, USA.

Hegrenes, A., B. Hill and G. Lien, 2001. Income instability among agricultural households – evidence from Norway. Journal of Farm Management, 11, 1-12.

Hirshleifer, J. and J.G. Riley, 1992. The Analytics of Uncertainty and Information. Cambridge University Press, Cambridge, UK.

Hueth, D.L. and W.H. Furtan (ed.), 1994. Economics of agricultural crop insurance: theory and evidence, natural resource management and policy. Kluwer Academic Publishers, Dordrecht, The Netherlands.

Just, R.E. and Q. Weninger, 1999. Are crop yields normally distributed? American Journal of Agricultural Economics, 81, 287-304.

Kapronczai, I. (ed.), 2005. A mezőgazdasági termelők alkalmzakodóképességének jellemzői. Agricultural Economics Research Institute, Hungary.

Kaylen, M.S., E.T. Loehman and P.V. Preckel, 1989. Farm-level analysis of agricultural insurance: a mathematical programming approach. Agricultural Systems, 30, 235-244.

Kennedy G. and Y. Yang, 1982. An analysis of alternative income stabilisation programs for British Columbia beef producers. Canadian Journal of Agricultural Economics, 30, 285-296.

KSH /Hungarian Central Statistical Office, 2004. Agriculture in Hungary – farm typology, 2000, 2003. KSH, Hungary.

Lehtonen, R. and E. Pahkinen, 2004. Practical methods for design and analysis of complex surveys. 2nd edition, John Wiley & Sons, UK.

Lianos, T.P. and D. Parliarou, 1986. Farm size structure in Greek agriculture. European Review of Agricultural Economics, 13, 233-248.

Lohr, S.A., 1999. Sampling: design and analysis. Duxbury Press, USA.

Majewski, E., W. Guba and A. Was, 2007. Farm income risk assessment for selected farm types in Poland – implications of future policy reforms. In: 16th congress of IFMA, 15th-20th July, 2007. Cork, Ireland.

Malhotra, N.K., 1999. Marketing research: an applied orientation. 3rd edition, Thomson Brooks/Cole, USA.

Mangen, M.-J.J. and A.M. Burrell, 2003. Who gains, who loses? Welfare effects of classical swine fever epidemics in the Netherlands. European Review of Agricultural Economics, 30, 125-154.

Markowitz, H., 1959. Portfolio selection – efficient diversification of investments. Wiley, New York, USA.

Meuwissen, M.P.M., R.B.M. Huirne and J.B. Hardaker, 1999. Income insurance in European agriculture. European Economy, Reports and Studies No. 2. Office for Official Publications of the European Communities, Luxembourg.

Meuwissen, M.P.M., R.B.M. Huirne and J.B. Hardaker, 2001. Risk and risk management: an empirical analysis of Dutch livestock farmers. Livestock Production Science, 69, 43-53.

Meuwissen, M.P.M., R.B.M. Huirne and J.R. Skees, 2003. Income insurance in European agriculture. EuroChoices, Spring 2003, 12-17.

Meyer, J., 1987. Two-moment decision models and expected utility maximization. American Economic Review, 77, 421-430.

Mishra, P.K., 1996. Agricultural risk, insurance and income. a study of in the impact and design of india's comprehensive crop insurance scheme. Avery, Aldershot, UK.

Mishra, A.K. and B.K. Goodwin, 1997. Farm income variability and the supply of off-farm labor. American Journal of Agricultural Economics, 79, 880-887.

Mishra, A.K. and B.K. Goodwin, 2003. Adoption of crop insurance versus revenue insurance: a farm-level analysis. Agricultural Finance Review, Fall, 144-155.

Musser, W.N. and G.F. Patrick, 2002. How much does risk really matter to farmers? In: Just and Pope (ed.), 'A Comprehensive Assessment of the Role of Risk in U.S. Agriculture'. Kluwer Academic Publishers, Norwell, 537-556.

Mußhoff, O. and N. Hirschauer, 2008. Hedging von Mengenrisiken in der Landwirtschaft – Wie teuer dürfen 'ineffective' Wetterderivate sein? Agrarwirtschaft (forthcoming).

Nakajima, C., 1986. Subjective equilibrium theory of the farm household. Elsevier, Amsterdam, The Netherlands.

Niemi, J. and J. Ahlstedt, 2007. Finnish Agriculture and Rural Studies 2007. MTT Agrifood Research Finland, Publications 107a, Helsinki, Finland.

OECD, 2000a. Income Risk Management in Agriculture. Organisation for Economic Co-operation and Development, Paris, France.

OECD, 2000b. Approaches to income risk management in OECD countries. Organisation for Economic Co-operation and Development, Paris, France.

OECD, 2002, Farm household income issues in OECD countries: A synthesis report. AGR/CA/APM(2002)11/FINAL, also published as OECD (2003) farm household income – issues and policy responses. ISBN 92-64-09965-4. Organisation for Economic Co-operation and Development, Paris, France.

OECD, 2004a. Farm household income: towards better informed policies. Policy Brief, OECD Observer. Organisation for Economic Co-operation and Development, Paris, France.

OECD, 2004b. Risk effects of PSE crop measures. Organisation for Economic Co-operation and Development, Paris, France.

OECD, 2005. Agricultural policies in OECD countries: monitoring and evaluation. Organisation for Economic Co-operation and Development, Paris, France.

OECD, 2007a. Agricultural policies in OECD countries: monitoring and evaluation 2007. Organisation for Economic Co-operation and Development, Paris, France.

OECD, 2007b. Commodity market impacts of trade and domestic agricultural policy reform. Organisation for Economic Co-operation and Development, Paris, France.

OECD-FAO, 2007. The OECD-FAO Agricultural Outlook 2007-2016. Organisation for Economic Co-operation and Development, Paris, France.

Offutt, S., 2002. The future of farm policy analysis: a household perspective. American Journal of Agricultural Economics, 84, 1229-37.

Ogurtsov, V.A., M.A.P.M. Van Asseldonk and R.B.M. Huirne, 2007. Insurance decisions by Dutch arable and dairy farmers. Journal of Farm Management, 13, 95-106.

Pannell, D.J., G. Hailu, A. Weersink and A. Burt, 2007. More reasons why farmers have so little interest in futures markets. Working Paper 0702, School of Agricultural and Resource Economics, Crawley, University of Western Australia,.

Phimister, E., 1993. Savings and investment in farm households: analysis using life cycle. Avebury, Aldershot, UK.

Phimister, E., D. Roberts and A. Gilbert, 2004. The dynamics of farm incomes: panel data analysis using the farm accounts survey. Journal of Agricultural Economics, 55, 197-220.

Polomé, P., O. Harmignie and B. Henry De Frahan, 2006. Farm-level acreage allocation under risk. In: 'American Agricultural Economics Association Annual Meeting, Long Beach, California, July 23-26'. California, USA

Poppe, K.J. and H. van Meijl, 2006. Differences in Farm Performance and Adjustment to Change: A Perspective from the Netherlands, in D. Blandford and B. Hill (eds.), 'Policy Reform and Adjustment in the Agricultural Sectors of Developed Countries'. Cabi Publishing.

Quiggin, J. and R.G. Chambers, 2004. Drought policy: a graphical analysis. Australian Journal of Agricultural and Resource Economics, 48, 225-251.

Ray, P.K., 1981. Agricultural insurance: theory and practice and application to developing countries, 2nd Edition. Pergamon Press, Oxford, UK.

Rejda, G.E., 1995. Principles of risk management and insurance. Harper Collins Publishers, New York, USA.

Richardson, J.W., G. Lien and J.B. Hardaker, 2006. Simulating multivariate distributions with sparse data: a kernel density smoothing procedure. In: 26th International Conference of the IAAE , August 12-18, 2006. Gold Coast, Australia.

RMA (Risk Management Association, U.S. Department of Agriculture), 2008. 'Group Risk Income Protection'. Crop Policies Webpage, http://www.rma.usda.gov/policies/GRIP.html

Robison, L.J. and P.J. Barry, 1987. The competitive firm's response to risk. Macmillan, New York, USA.

Robison, L.J. and S.D. Hanson, 1997. Analysing firm response to risk using mean-variance models. In: R.B.M. Huirne, J.B. Hardaker and A.A Dijkhuizen (ed), 'Risk management strategies in agriculture: state of the art and future perspectives, 121-137'. Mansholt Institute, Wageningen, The Netherlands.

Schmitz, B., 2007. Wetterderivate als Instrument des Risikomanagements landwirtschaftlicher Betriebe. Dissertation, Bonn, Germany. Online: http://hss.ulb.uni-bonn.de/diss_online/landw_fak/2007/schmitz_bernhard.

Sherrick, B.J., P.J. Barry, P.N. Ellinger and G.D. Schnitkey, 2004. Factors influencing farmers' crop insurance decisions. American Journal of Agricultural Economics, 86, 103-114.

Skees, J.R., 1999. Agricultural risk management or income enhancement? Regulation: The Cato Review of Business and Government, 22, 35-43.

Skees, J.R., 2001. The bad harvest: crop insurance reform has become a good idea gone awry. Regulation, 24, 16-21.

Skees, J.R., 2007. Challenges for use of index-based weather insurance in lower income countries. In: 'Management of climate risks in agriculture', 101st EAAE Seminar, July 5-6, 2007. Berlin, Germany.

Starp, M., 2006. Integriertes Risikomanagement im landwirtschaftlichen Betrieb. Betriebswirtschaftliche Schriften, Heft 166. Duncker and Humblot, Berlin, Germany.

Stokes, J.R., K.H. Coble and R. Dismukes, 2000. Producer behaviour in the presence of an income stabilisation program. Agricultural Finance Review, 60, 33-59.

Swinbank, A., 2005a. Developments in the WTO and implications for CAP. In: 'Agricultural Economics Society's one-day conference, CAP Reformed?, 12 January 2005'. London, UK

Swinbank, A., 2005b. The evolving CAP, pressures for reform, and implications for trade policy. In: 'Australian Agricultural and Resource Economics Society's pre-conference workshop: Trade Policy Reform and Agriculture: Prospects, Strategies, Implications, Coffs Harbour, NSW, 8 February 2005'. New South Wales, Australia.

Torriani, D., P. Calanca, M. Beniston and J. Fuhrer, 2007. Alternative hedging strategies in maize production to cope with climate variability and change. In: 'Management of climate risks in agriculture', 101st EAAE Seminar, July 5-6, 2007. Berlin, Germany.

Turvey, C., 2007. Weather risk contingent credit instruments. In: 'Management of climate risks in agriculture', 101st EAAE Seminar, July 5-6, 2007. Berlin, Germany.

Van Asseldonk, M.A.P.M., M.P.M. Meuwissen and R.B.M. Huirne, 2002. Belief in disaster relief and demand for a public-private insurance program. Review of Agricultural Economics, 24, 196-207.

Van Asseldonk, M.A.P.M., O. Kobzar, M.P.M. Meuwissen, R.B.M. Huirne and J.B. Hardaker, 2006. Market-based crop insurance appraisal using whole-farm planning. Income stabilisation in agriculture. The role of public policy. Editors: C. Cafiero and A. Cioffi, 409-417. Edizioni Scientifiche Italine, Napoli, 2006.

Williams, C.A., M.L. Smith and P.C. Young, 1995. Risk management and insurance. 7$^{th}$ edition, McGraw-Hill, USA.

# Appendix A

## Reflections from an Eastern EU perspective

Income Stabilisation project findings were elaborated on during a one-day workshop near the end of the project. As the workshop's focus was on Eastern EU agricultural circumstances, the workshop was organised in Warsaw, in order to facilitate local experts to participate. The workshop was attended by 26 experts, originating from both Eastern EU member states and non-Eastern EU member states. Experts evaluated statements based on the Income Stabilisation project and assessed their applicability to Eastern EU agricultural circumstances. Evaluations were on a four-point scale, ranging from 1 (fully disagree) to 4 (fully agree). Results are shown in Table A.1. Statements exactly reflecting Income Stabilisation project conclusions are marked with an asterisk.

Table A.1 shows that experts generally agree on project conclusions to be applicable to Eastern EU agriculture. Strongest agreement, both for experts originating from Eastern EU member states and non-Eastern EU members states, was found for the need for public policy to facilitate private markets by educating farmers and extension workers in risk management issues and the use of derivate markets (statement 27). Agreement scores for both groups of experts are 3.33 and 3.42 respectively. Experts also agree on crises to be mainly caused by weather and market risks (statement 14). They furthermore agree that some member states are more exposed to income falls than others (statement 7) and (as a consequence) that risk premiums per hectare differ substantially across member states (statement 20). There is also agreement on decoupled direct payments playing a key role in stabilising farm incomes under a less protective CAP (statement 6). Experts in addition agree that many innovations, such as derivatives and public-private risk sharing, enlarge the opportunities for transferring risks, although experts originating from non-Eastern EU member states seem to be more confident about this issue.

Table A.1 also shows that there is some disagreement about the general applicability of some of the project conclusions to Eastern EU agriculture. This is for instance the case with regard to the expected farm-level impact of future scenarios. Experts, especially those not originating from Eastern EU member states, do expect a significant impact from WTO agreements on farm incomes (statement 5). In addition, they believe that more liberal policies actually will induce arable farmers in Eastern EU member states to change their farm plans (statement 18).

**Table A.1**. *Agreement on Income Stabilisation issues from the perspective of Eastern EU agricultural circumstances (1=fully disagree, 2=disagree, 3=agree, 4=fully agree). More than 50% 'agreements' are in bold (n=26)[1].*

| Statements[2] | 1 (%) | 2 (%) | 3 (%) | 4 (%) | Mean scores[3] Eastern EU (n=12) | Mean scores[3] Other EU (n=12) |
|---|---|---|---|---|---|---|
| *Past risk exposure* | | | | | | |
| 1 Income fluctuations are mainly determined by yield risks | 4 | 54 | 42 | - | 2.58 | 2.25 |
| 2 Farmers are used to large income fluctuations* | 4 | 3 | **58** | **8** | 2.75 | 2.67 |
| 3 Catastrophic events hardly cause farmers to go broke* | 8 | 31 | **50** | **11** | 2.83 | 2.58 |
| 4 FADN data can not be used to trigger (crisis) risk payments* | - | 23 | **46** | **31** | 2.92 | 3.17 |
| *Future risk exposure* | | | | | | |
| 5 Overall, new WTO agreements have no significant impact on farm income levels and risk of low incomes* | 19 | 54 | 23 | 4 | $2.42^{**}$ | $1.83^{**}$ |
| 6 In a less protective CAP, decoupled direct payments play a key role in stabilising farm incomes* | - | 12 | **65** | **23** | 2.91 | 3.25 |
| 7 Some member states are more exposed to income falls than others* | - | 4 | **65** | **31** | 3.17 | 3.33 |
| 8 Crop farms face higher prob. of neg. income than livestock farms* | 4 | 46 | 46 | 4 | 2.67 | 2.25 |
| 9 Small (< 16 esu) and large farms (> 100 esu) are more threatened by the risk of low incomes than mid-sized farms* | 4 | 54 | 38 | 4 | 2.55 | 2.27 |
| *Review of the risk management arena* | | | | | | |
| 10 The performance of publicly provided crop insurance is improved due to e.g. surveillance and better risk evaluations* | - | 25 | **67** | **8** | 2.82 | 2.91 |
| 11 Governments increasingly require farmers to contract insurance for being eligible for ad hoc disaster payments* | - | 16 | **60** | **24** | 3.00 | 3.08 |
| 12 Many innovations, such as derivatives and public-private risk sharing, enlarge the opportunities for transferring risks* | - | 12 | **56** | **32** | $3.00^{**}$ | $3.5^{**}$ |
| 13 Public(-private) compensation schemes for contagious animal diseases face problems of moral hazard* | - | 29 | **63** | **8** | 2.80 | 2.83 |

| | | | | | | |
|---|---|---|---|---|---|---|
| *Risk and risk management perception* | | | | | | |
| 14 Crises are mainly caused by weather and market risks* | - | 19 | **46** | **35** | 3.25 | 3.17 |
| 15 Property insurance is the most important risk management tool | - | 42 | **50** | **8** | 2.75 | 2.58 |
| 16 Farmers are fully aware of all risk management tools available | 27 | 62 | 11 | - | 1.83 | 1.92 |
| 17 Decoupled payments are not perceived as a risk management tool | 12 | 19 | **61** | **8** | 2.50 | 2.75 |
| *The economics of risk management instruments* | | | | | | |
| 18 More liberal policies do not induce arable farmers to change their farm plan* | 4 | 67 | 25 | 4 | 2.20 | 2.42 |
| 19 Diversification is not likely to become a key risk management tool* | 8 | 42 | 42 | 8 | 2.58 | 2.33 |
| 20 Risk premiums per hectare differ substantially across member states* | - | 4 | **71** | **25** | 3.08 | 3.40 |
| 21 Insurance schemes are only attractive for catastrophic events | 15 | 66 | 19 | - | 2.42* | 1.67* |
| 22 Insurance schemes are only attractive in case of premium subsidies | 4 | 52 | 44 | - | 2.33 | 2.36 |
| *Policy options for risk management* | | | | | | |
| 23 Crisis risks are unforeseen, happen infrequently and related losses exceed the individual capacity to cope* | - | 12 | **61** | **27** | 3.08 | 3.25 |
| 24 The only short-term way for public policy to deal with crisis risk, is to provide direct damage compensation* | 8 | 39 | **38** | **15** | 2.58 | 2.58 |
| 25 Conditions for providing disaster relief need to be set at EU level* | 4 | 35 | **42** | **19** | 2.75 | 2.83 |
| 26 Insurance schemes for production risks are underdeveloped | - | 19 | **73** | **8** | 2.75 | 3.00 |
| Public policy needs to facilitate private markets by: | | | | | | |
| 27 Educating farmers and extension workers in risk management issues and the use of derivate markets* | - | - | **62** | **38** | 3.33 | 3.42 |
| 28 Supporting the development of insurance and derivative markets, but *without* premium subsidies* | 8 | 38 | **35** | **19** | 2.58 | 2.75 |
| 29 Developing and operating mutual funds for specialty crops and animal diseases* | 8 | 15 | **54** | **23** | 3.08 | 2.75 |
| 30 Developing risk management services that take over farmers' complex risk management tasks* | 4 | 19 | **65** | **11** | 3.00 | 2.75 |

[1]12 experts are from Eastern EU member states, 12 from other EU member states and 2 responded anonymously.

[2]Asterisks (*) reflect Income Stabilisation project conclusions.

[3]Asterisks (* and **) indicate significant differences, i.e. at $p \leq 0.005$ and $p \leq 0.10$ respectively.

With regard to the farm types and size classes more likely to face income risks, experts are not convinced that in Eastern EU circumstances these are crop farms (statement 8) and small and large farms (statement 9). Furthermore, 50% regards diversification as potentially becoming an important risk management tool in Eastern European agriculture (statement 19).

With regard to the statements not directly based on Income Stabilisation project conclusions, experts generally agree that farmers are not fully aware of the wide spectrum of risk management tools available (statement 16). From all statements given, experts originating from non-Eastern EU member scored their most extreme 'disagreement score' (i.e. 1.83) on this statement, probably reflecting their strong and very similar opinions about this issue. Experts also agree that insurance schemes for production risks are underdeveloped (statement 26) and that insurance schemes are not only attractive in case of subsidies (statement 22) and catastrophic events (statement 21). The wider applicability of insurance schemes (statement 21) was especially stressed by experts from non-Eastern EU member states.

# Appendix B

## Assessment of policy scenarios

The base model for simulation of policy scenario effect on farm income distribution

The level and volatility of farm incomes were estimated using a Monte Carlo simulation method in a farm model constructed for the @Risk package. The key parameters of the base model which were calculated from historical data can be grouped as follows:
- means of structural variables to describe the farm types (e.g. size of activities, yield, prices, inputs or costs) calculated from FADN data base for the years 2002-2004;
- standard Deviation for selected variables;
- cross correlations: (1) farm related (input-output, input-input) from historical farm data; (2) market related (price-price, price-yield; yield-yield) from national statistics data.

Due to data limitations input-output correlations for crop production were not included in the model. Most of the farm activities in the model were described by the parameters of the distributions (standard deviation) of yields and prices. Similarly, the standard deviation was estimated for selected cost variables (energy, fertilizers, pesticides, seeds, purchased and farm produced feed for animals). Other variables of the model (e.g. fixed costs) were introduced as constant values specific for each farm type. For simplification a normal distribution for all variables was assumed. The distribution was truncated on the left side at 0 for yields and for prices at the values, optionally, of $\bar{x} - 2\sigma$ or 0 or the intervention price, depending on which was the highest.

The estimation of standard deviation in the base period, which is a basic measure of instability of yields and prices in the simulation model, created some difficulties related mainly to available sources of data. In the case of Poland data from two different sources have been merged: FADN results for the period 2002-2004 and Farm Survey[10] for the years 1997-2001, adjusted to FADN standards. For a given farm type (activity, size) all observations have been pooled across years (1997-

[10] Farm Survey conducted by the Institute of Agricultural and Food Economics in Warsaw. Polish FADN, which have been established very recently, provides data for the years 2002-2004 only, but for a large sample of farms (12,000 in the year 2004).

2004) and standard deviations were estimated for the whole set of variables. Because the Farm Survey, which is not fully compatible with FADN, provides historical data for a long period, however for much smaller population of farms (about 1000 on average in the period considered), all farms from the Farm Survey which represent farm types selected for simulations were merged with randomly drawn 10% of FADN farm population.

Splitting the population of farms into selected farm types, and drawing data on single activities from smaller samples, which do not appear in all farms, reduced strongly the number of observations which can be used for estimation. That is why it was decided to pool all the observations within each farm type and estimate the standard deviation for the whole set of variables. Consequently, the analysis and its results are interpreted in relation to the experienced (*ex-post* for the base period) and envisaged (*ex-ante* for scenario analysis) situation in the population of farms, rather then in a single farm.

Any estimates of means and standard deviations from the pooled data in the simulations produce a randomly chosen value that depends on all the combined sources of variation, including the hopefully small net sampling errors, between farm performance levels and year to year variations due to weather and market/policy conditions.

As a result, the simulated income distributions for the represented farm population show the proportion of all farms likely to fall below some critical level, reflecting their economic viability. The statistics (mean and SD) capture all the variation even though it is not separated out into its respective components.

**Table B.1.** *Determining price level assumptions for policy scenarios – summary table.*

| Base (2002-2004) | Most likely (2013) | Likely A (2018) | Likely B (2018) | Liberal (2018) | Protectionist (2018) |
|---|---|---|---|---|---|
| Statistics and FADN | OECD-FAO Projection 2007-2016 | OECD-FAO Projection 2007-2016 (extended to 2018) | Range of assumed price decreases (0-25%) | Range of assumed price decreases (5-35%) | 5-10% above 2016 projections (except for sugar, potatoes) |
| | | | Expected reduction of prices resulting from policy liberalisation<br>Stronger for commodities with higher initial price support (measured by nominal protection coefficient; OECD, PSE database)<br>Price reductions moderated by a positive impact of liberalisation on world market prices | | |
| | | | Own assumptions based on OECD-FAO projections | | |

Sources: The OECD-FAO Agricultural Outlook 2007-2016; OECD: Agricultural Policies in OECD Countries: Monitoring and Evaluation 2007; U.S. Proposal for WTO Agriculture Negotiations: Its Impact on U.S. and World Agriculture, FAPRI; CARD Working Paper 05-WP 417, December 2005; OECD (2007) Commodity Market Impacts of Trade and Domestic Agricultural Policy Reform, 2007.

**Table B.2.** *World market prices as % of the EU prices (estimates of nominal protection from OECD PSE calculation).*

| Product | 1986 - 88 | 2004 -2006 | 2004 | 2005 | 2006p |
|---|---|---|---|---|---|
| Sugar | 29 | 40 | 32 | 38 | 57 |
| Beef and veal | 44 | 51 | 47 | 50 | 56 |
| Poultry | 56 | 57 | 52 | 60 | 61 |
| Sheep meat | 35 | 64 | 68 | 65 | 58 |
| Milk | 21 | 71 | 50 | 100 | 75 |
| Pig meat | 78 | 81 | 75 | 85 | 86 |
| Maize | 45 | 88 | 83 | 85 | 97 |
| Other grains | 41 | 99 | 98 | 98 | 100 |
| Wheat | 47 | 100 | 100 | 99 | 100 |
| Rapeseed | 42 | 100 | 100 | 100 | 100 |
| Eggs | 81 | 100 | 100 | 100 | 100 |

OECD: Agricultural Policies in OECD Countries : Monitoring and Evaluation 2007 (and own calculations).

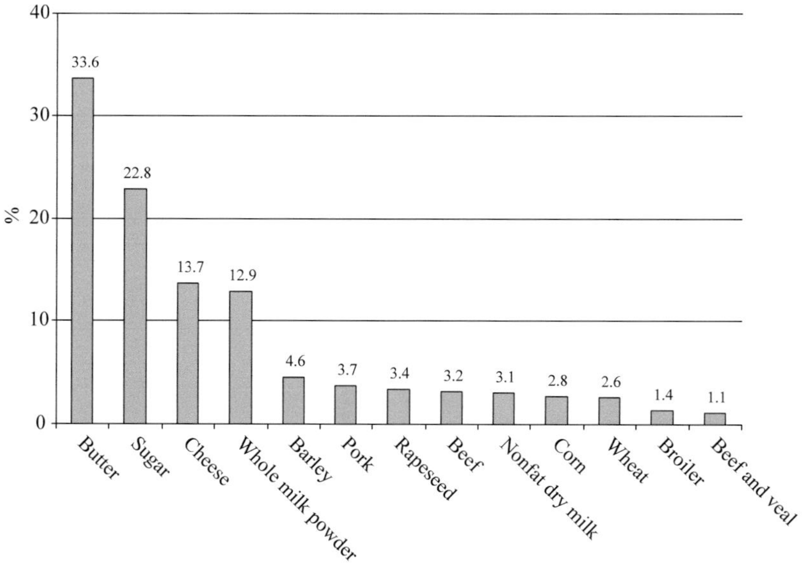


**Figure B.1.** *Percentage changes in world prices – Impact of U.S. Proposal for WTO Negotiations. Source: U.S. Proposal for WTO Agriculture Negotiations: Its Impact on U.S. and World Agriculture, FAPRI, CARD Working Paper 05-WP 417, December 2005.*

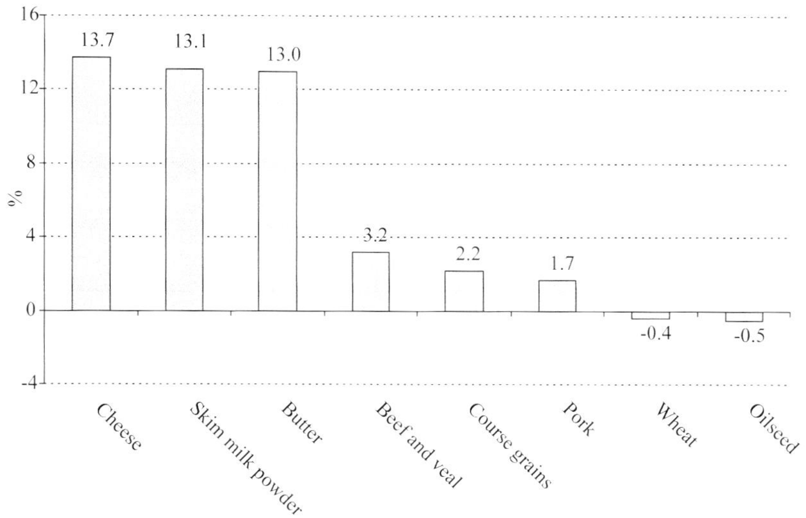


**Figure B.2.** *Percentage changes in world prices – Impact of 50% liberalisation. Source: OECD Commodity Market Impacts of Trade and Domestic Agricultural Policy Reform, 2007.*

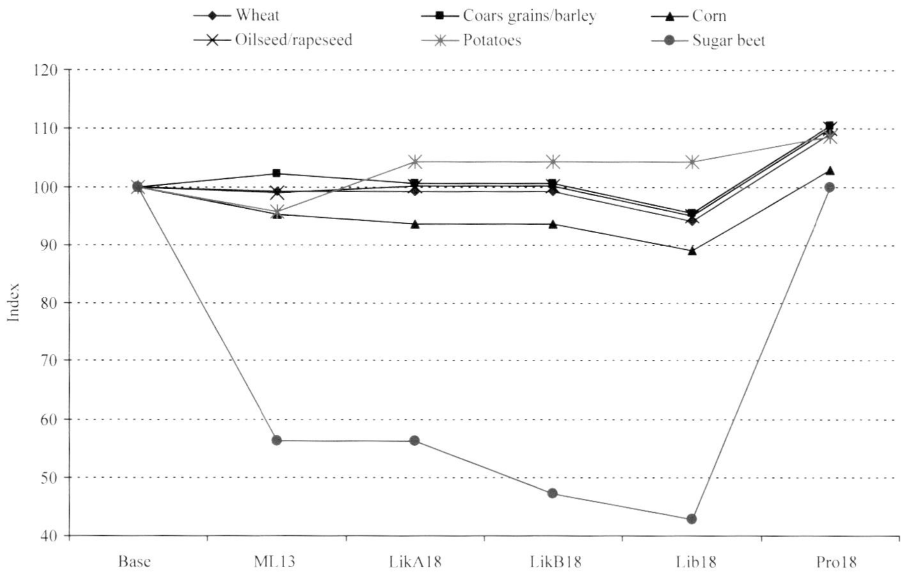


**Figure B.3.** *Price level assumption for crops for policy scenarios, Base (2002-2004) = 100. Source: own assumptions based on methodology described in the text.*

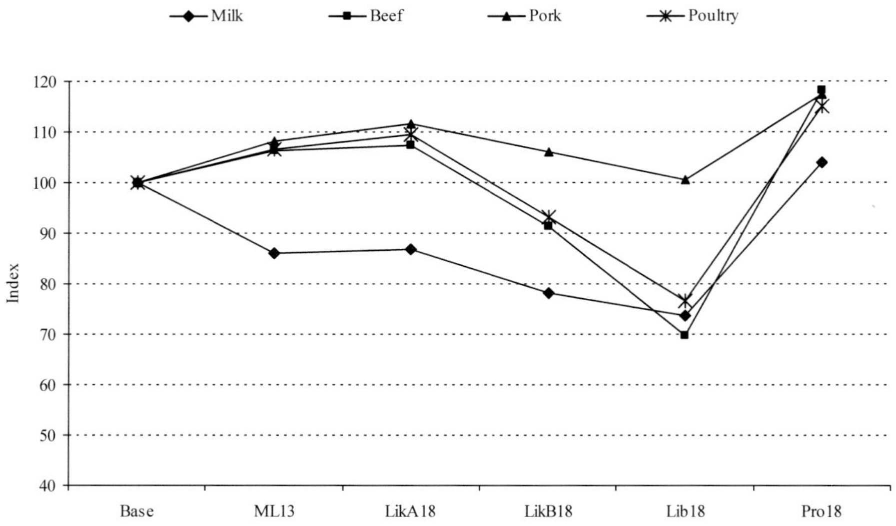


**Figure B.4.** *Price level assumptions for animal products for policy scenarios, Base (2002-2004) = 100. Source: own assumptions based on methodology described in the text.*

**Table B.3.** *Volatility of agricultural prices in EU and World market - Coefficient of Variation for period 1993-2005.*

| Product | World market | EU established member states | CV (%) | EU new member states | CV (%) (1996-2003) |
|---|---|---|---|---|---|
| Wheat | 19.3 | Denmark | 9.5 | Hungary | 19.9 |
| | | France | 11.4 | Poland | 9.8 |
| | | Germany | 12.4 | Averages[1]: | |
| | | Netherlands | 12.3 | EMS | 11.0 |
| | | Spain | 11.4 | NMS | 14.9 |
| | | UK | 9.1 | | |
| Potatoes | n.a. | Denmark | 25.4 | Hungary | 33.6 |
| | | France | 39.6 | Poland | 20.2 |
| | | Germany | 50.2 | Averages[1]: | |
| | | Netherlands | 45.4 | EMS | 35.9 |
| | | Spain | 19.8 | NMS | 26.9 |
| | | UK | 34.9 | | |
| Sugar beet | 23.5 (sugar) | Denmark | 6.8 | Hungary | 26.3 |
| | | France | 4.8 | Poland | 10.5 |

**Table B.3.** *Continued.*

| Product | World market | EU established member states | CV (%) | EU new member states | CV (%) (1996-2003) |
|---|---|---|---|---|---|
| Sugar beet (Continued) | | Germany | n.a. | Averages[1]: | |
| | | Netherlands | 11.2 | EMS | 8.0 |
| | | Spain | n.a. | NMS | 18.4 |
| | | UK | 9.1 | | |
| Raw milk | n.a. | Denmark | 1.8 | Hungary | 23.1 |
| | | France | 2.0 | Poland | 15.0 |
| | | Germany | 4.4 | Averages[1]: | |
| | | Netherlands | 4.8 | EMS | 4.0 |
| | | Spain | 4.8 | NMS | 19.1 |
| | | UK | 6.1 | | |
| SMP | 16.2 | Denmark | - | Hungary | n.a. |
| | | France | 7.3 | Poland | n.a. |
| | | Germany | - | Averages[1]: | |
| | | Netherlands | 6.8 | EMS | 7.6 |
| | | Spain | 6.7 | NMS | n.a. |
| | | UK | 9.5 | | |
| Cattle | 10.8 | Denmark | 1.6 | Hungary | 12.1 |
| | | France | 8.1 | Poland | 4.9 |
| | | Germany | n.a. | Averages[1]: | |
| | | Netherlands | 16.7 | EMS | 8.7 |
| | | Spain | 8.2 | NMS | 8.5 |
| | | UK | n.a. | | |
| Pork | 17.3 | Denmark | 14.7 | Hungary | 21.3 |
| | | France | - | Poland | 11.1 |
| | | Germany | - | Averages[1]: | |
| | | Netherlands | 14.3 | EMS | 13.6 |
| | | Spain | 13.1 | NMS | 16.2 |
| | | UK | 12.3 | | |
| Poultry | 8.9 | Denmark | 4.9 | Hungary | 9.4 |
| | | France | 4.1 | Poland | 9.0 |
| | | Germany | 5.5 | Averages[1]: | |
| | | Netherlands | 7.6 | EMS | 6.3 |
| | | Spain | 9.5 | NMS | 9.2 |
| | | UK | - | | |

[1]EMS=established member states; NMS=new member states.

Source: own estimates based on FAO database.

**Table B.4.** *Changes in volatility of world market prices due to policy change in EU and other countries.*

| | Wheat | Coarse grains | Oilseed | Rice |
|---|---|---|---|---|
| Observed data 1986-2001: CV with partial price transmission in all countries | 0.25 | 0.19 | 0.17 | 0.25 |
| CV with complete price transmission only in the EU | 0.16 | 0.15 | 0.15 | 0.23 |
| CV with complete price transmission the EU Switzerland, Japan, Canada, Mexico, United States | 0.14 | 0.13 | 0.13 | 0.2 |
| Reduction in variability when allowing complete price transmission | 45% | 32% | 23% | 21% |

Source: OECD (2004) Risk effects of PSE crop measure.

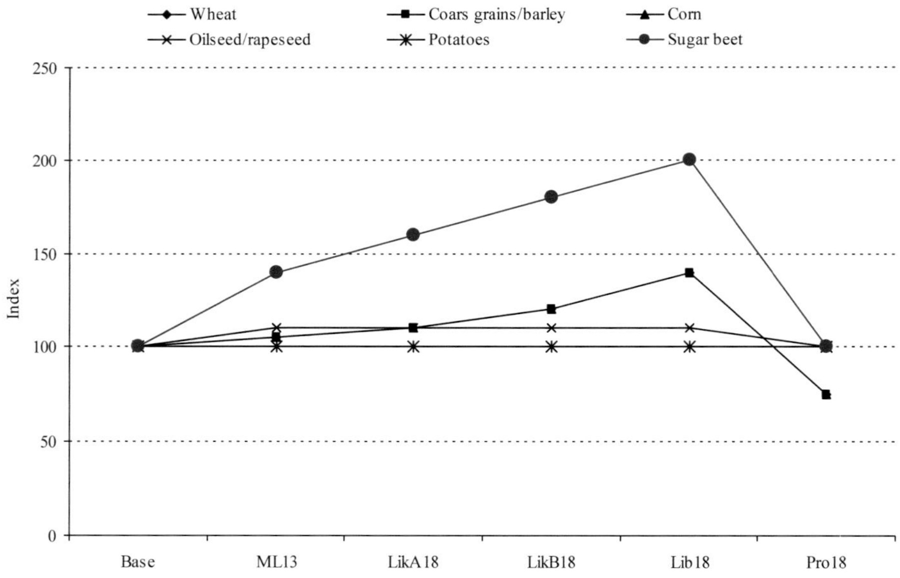


**Figure B.5** *Assumptions for volatility of crop prices for policy scenarios, Base (2002-2004) = 100. Source: own assumptions based on methodology described in the text.*

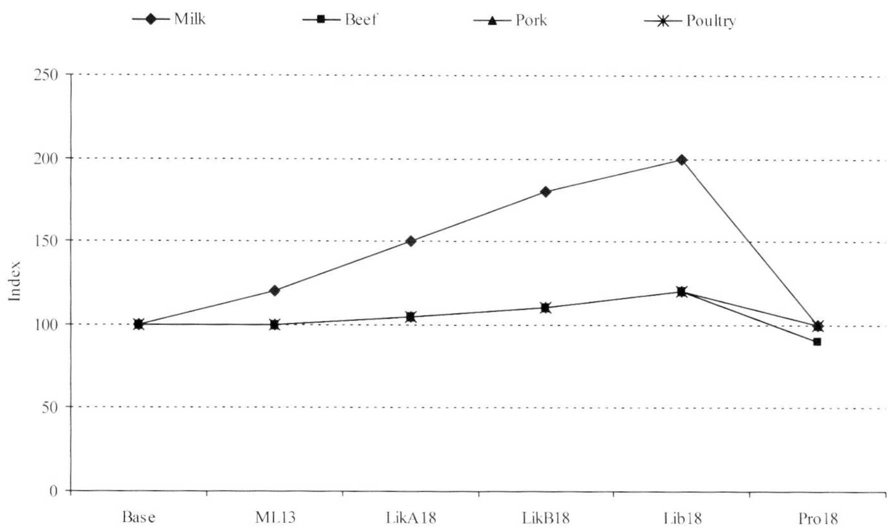


**Figure B.6.** *Assumptions for volatility of prices of animal products for policy scenarios, Base (2002-2004) = 100. Source: own assumptions based on methodology described in the text.*

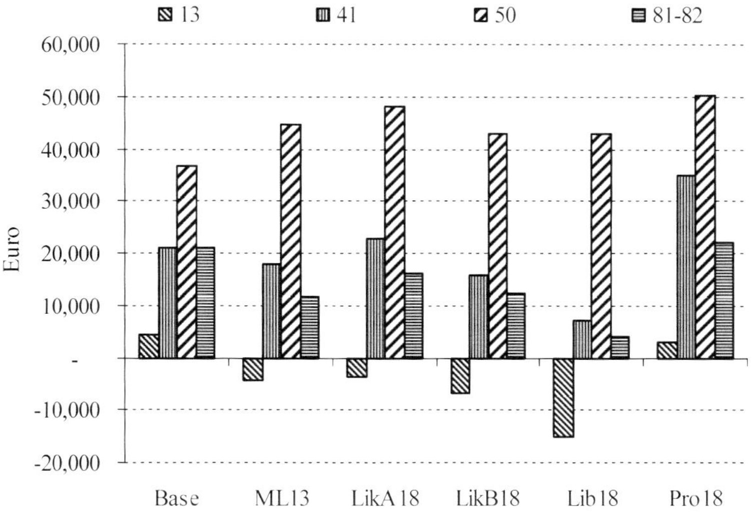


**Figure B.7A.** *Mean farm incomes under different policy scenarios in selected FADN farm types in Germany (size class: 16-40 ESU; type 13: arable; 41: dairy; 50: pigs; 81-82: mixed).*

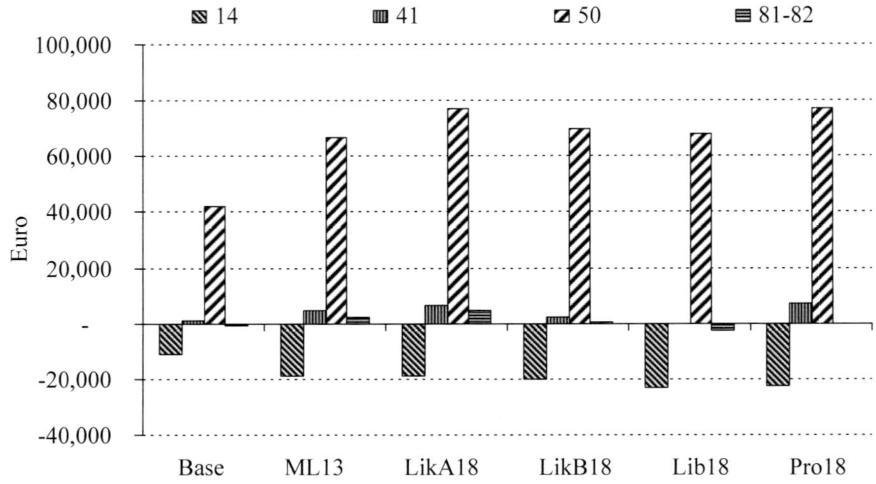


**Figure B.7B.** *Mean farm incomes under different policy scenarios in selected FADN farm types in the Netherlands (size class: 16-40 ESU; type 14: arable; 41: dairy; 50: pigs; 81-82: mixed).*

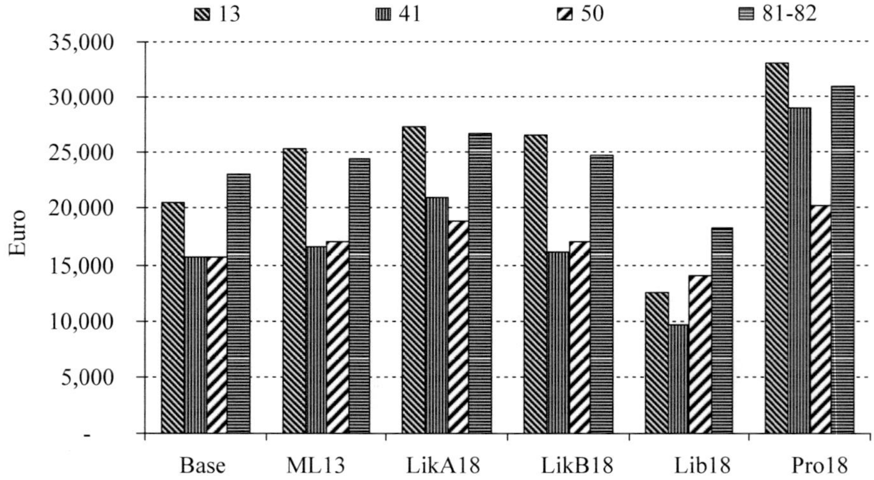


**Figure B.7C** *Mean farm incomes under different policy scenarios in selected FADN farm types in Poland (size class: 16-40 ESU; type 13: arable; 41: dairy; 50: pigs; 81-82: mixed).*

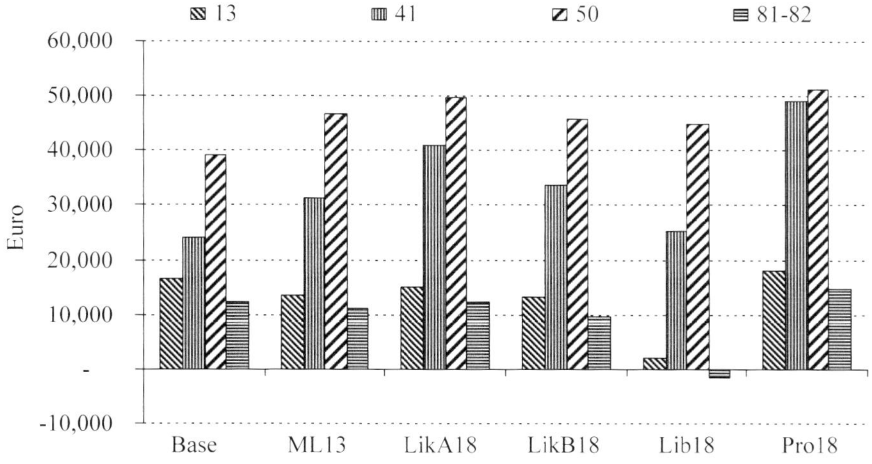


**Figure B.7D** *Mean farm incomes under different policy scenarios in selected FADN farm types in Spain (size class: 16-40 ESU; type 13: arable; 41: dairy; 50: pigs; 81-82: mixed).*

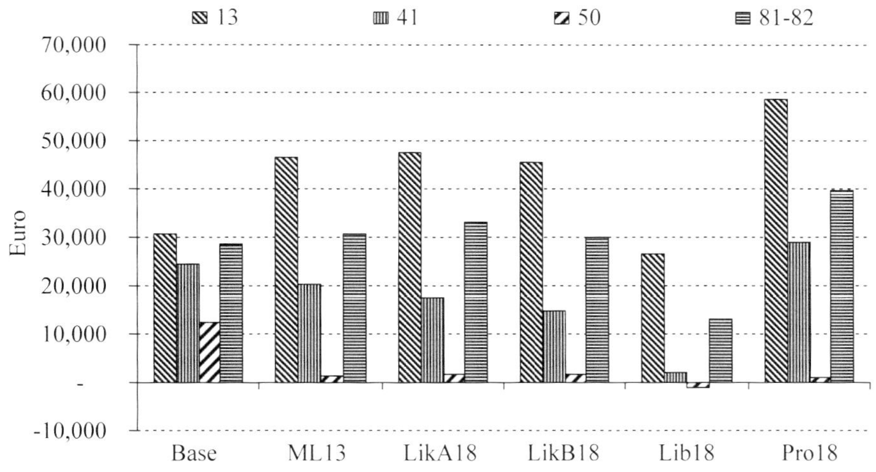


**Figure B.7E.** *Mean farm incomes under different policy scenarios in selected FADN farm types in Hungary (size class: 16-40 ESU; type 13: arable; 41: dairy; 50: pigs; 81-82: mixed).*

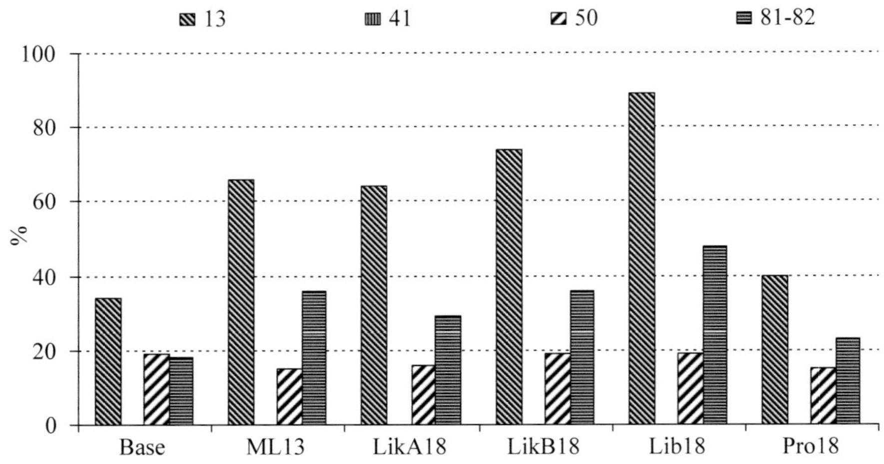


**Figure B.8A.** *Risk of low incomes under different policy scenarios in selected FADN farm types in Germany (size class: 16-40 ESU; type 13: arable; 41: dairy; 50: pigs; 81-82: mixed).*

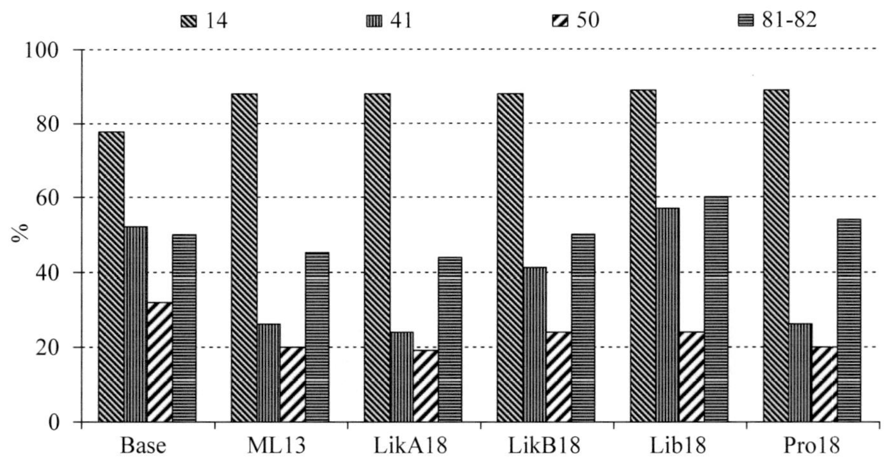


**Figure B.8B.** *Risk of low incomes under different policy scenarios in selected FADN farm types in the Netherlands (size class: 16-40 ESU; type 14: arable; 41: dairy; 50: pigs; 81-82: mixed).*

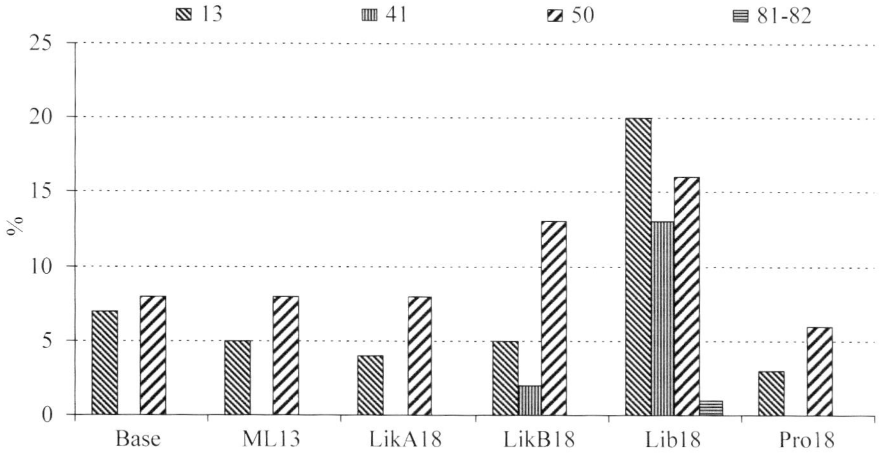


**Figure B.8C.** *Risk of low incomes under different policy scenarios in selected FADN farm types in Poland (size class: 16-40 ESU; type 13: arable; 41: dairy; 50: pigs; 81-82: mixed).*

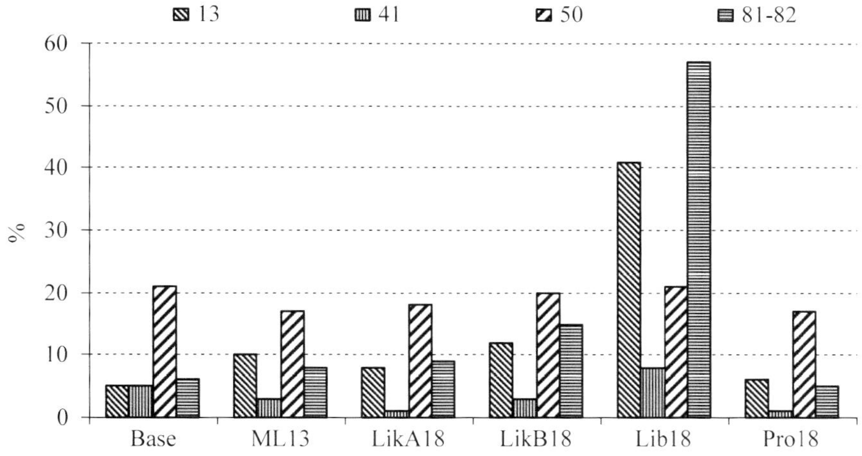


**Figure B.8D.** *Risk of low incomes under different policy scenarios in selected FADN farm types in Spain (size class: 16-40 ESU; type 13: arable; 41: dairy; 50: pigs; 81-82: mixed).*

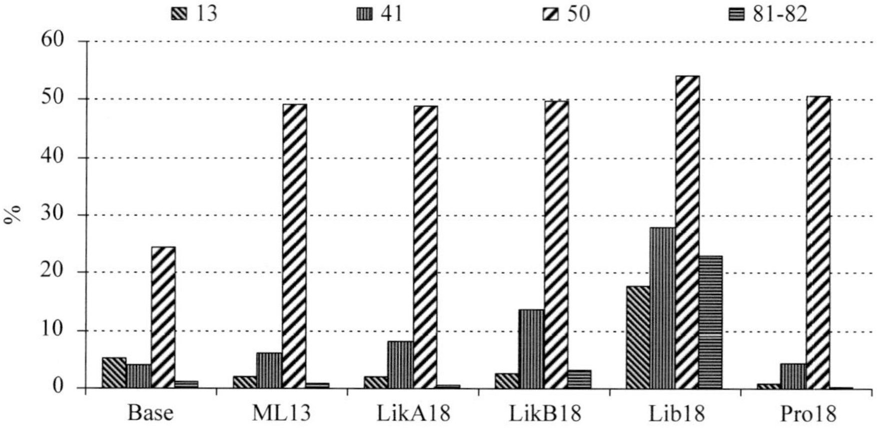


**Figure B.8E.** *Risk of low incomes under different policy scenarios in selected FADN farm types in Hungary (size class: 16-40 ESU; type 13: arable; 41: dairy; 50: pigs; 81-82: mixed).*

# Appendix C

## Descriptive statistics farmers' perceptions

**Table C.1a.** *Has the farmer ever had to face any unexpected event that threatened the farm with bankruptcy?*

| | Germany (A); n=201 | Hungary (B); n=201 | Netherlands (C); n=235 | Poland (D); n=206 | Spain (E); n=200 |
|---|---|---|---|---|---|
| Has any unexpected event threatening the farm ever occurred? | | | | | |
| Yes | 21.9% (-) | 40.3% (ACD) | 25.5% (-) | 17.5% (-) | 56.5% (ABCD) |
| No | 78.1% (BE) | 59.7% (E) | 74.5% (BE) | 82.5% (BE) | 43.5% (-) |
| Total | 100.0% | 100.0% | 100.0% | 100.0% | 100.0% |

**Table C.1b.** *Type of critical event (if there was any).*

| | Germany (A); n=201 | Hungary (B); n=201 | Netherlands (C); n=235 | Poland (D); n=206 | Spain (E); n=200 |
|---|---|---|---|---|---|
| Type of event leading to crisis | | | | | |
| Climatic | 23.3% (-) | 71.6% (AC) | 30.0% (-) | 54.3% (A) | 82.3% (ACD) |
| Epidemic | 4.7% (-) | 11.1% (-) | 30.0% (ABE) | 5.7% (-) | 5.3% (-) |
| Farmer's personal health | 14.0% (E) | 6.2% (-) | 26.7% (BE) | 31.4% (BE) | 1.8% (-) |
| Market conditions | 39.5% (E) | 50.6% (E) | 31.7% (E) | 31.4% (E) | 9.7% (-) |
| Policy measures | 25.6% (E) | 23.5% (E) | 26.7% (E) | 5.7% (-) | 3.5% (-) |

**Table C.2.** *Aspects of crisis situation in crop production in the last ten years.*

| | | Number of times unexpected yield loss exceeded 10% of planned yield in the last ten years | % of land affected by most critical yield loss (if such happened) | % of total farm revenue affected by the most critical yield loss (if such happened) |
|---|---|---|---|---|
| Germany (A) | Mean | 2.9 | 40.02 | 17.95 |
| | N | 192 | 164 | 164 |
| | Minimum | 0 | 0 | 0 |
| | Maximum | 10 | 100 | 100 |
| | SD | 1.8 | 27.38 | 15.13 |
| | Greater than | - | CD | - |
| Hungary (B) | Mean | 2.99 | 47.08 | 36.09 |
| | N | 192 | 173 | 171 |
| | Minimum | 0 | 2 | 0 |
| | Maximum | 10 | 100 | 100 |
| | SD | 1.95 | 28.44 | 20.7 |
| | Greater than | - | CD | AD |
| Netherlands (C) | Mean | 2.88 | 27.39 | 29.45 |
| | N | 82 | 36 | 31 |
| | Minimum | 0 | 0 | 0 |
| | Maximum | 10 | 100 | 100 |
| | SD | 2.23 | 30.01 | 28.16 |
| | Greater than | - | D | - |
| Poland (D) | Mean | 3.03 | 4.42 | 25.7 |
| | N | 204 | 177 | 175 |
| | Minimum | 0 | 0 | 0 |
| | Maximum | 50 | 90 | 80 |
| | SD | 3.8 | 12.81 | 15.38 |
| | Greater than | - | - | A |
| Spain (E) | Mean | 4.21 | 69.59 | 65.59 |
| | N | 149 | 143 | 143 |
| | Minimum | 0 | 33 | 33 |
| | Maximum | 10 | 100 | 100 |
| | SD | 2.27 | 24.41 | 26.12 |
| | Greater than | ABCD | ABCD | ABCD |

**Table C.3.** *Farmers' ratings of factors harmful to crop production (member state averages; 1-3- Harmless, 3-5-Moderately harmful, 5-7-Very harmful).*

| | | Viral and bacterial diseases | Fungi | Weed | Insects | Invertebrata | Vertebrata |
|---|---|---|---|---|---|---|---|
| Germany (A) | Mean | 5.82 | 5.6 | 4.68 | 4.77 | 4.21 | 3.87 |
| | N | 193 | 194 | 192 | 193 | 190 | 190 |
| | SD | 1.28 | 1.14 | 1.28 | 1.47 | 1.59 | 1.64 |
| | Greater than | BCD | D | D | D | BE | B |
| Hungary (B) | Mean | 5.04 | 5.45 | 4.95 | 5.17 | 2.43 | 3.11 |
| | N | 192 | 194 | 193 | 195 | 184 | 190 |
| | SD | 1.64 | 1.56 | 1.58 | 1.48 | 1.57 | 1.81 |
| | Greater than | D | D | DE | ADE | - | - |
| Netherlands (C) | Mean | 4.85 | 5.21 | 4.37 | 5.22 | - | - |
| | N | 85 | 87 | 84 | 89 | - | - |
| | SD | 1.76 | 1.64 | 1.82 | 1.54 | - | - |
| | Greater than | D | D | D | DE | - | - |
| Poland (D) | MMean | 3.05 | 2.8 | 3.07 | 3.66 | 5.23 | 4.9 |
| | N | 202 | 200 | 201 | 202 | 201 | 201 |
| | SD | 1.87 | 1.72 | 1.88 | 2 | 1.99 | 1.98 |
| | Greater than | - | - | - | - | ABE | ABE |
| Spain (E) | Mean | 5.68 | 5.2 | 4.31 | 4.43 | 3.23 | 3.86 |
| | N | 148 | 143 | 143 | 143 | 144 | 146 |
| | SD | 1.47 | 1.65 | 1.84 | 1.8 | 2.02 | 2 |
| | Greater than | BCD | D | D | D | B | B |

**Table C.4.** *Aspects of crisis situation in livestock production in the last ten years.*

| | | Number of times unexpected production losses exceeded 5% of planned livestock production in the last ten years | % of livestock affected by most critical production loss (if such happened) | % of total farm revenue affected by the most critical production loss (if such happened) |
|---|---|---|---|---|
| Germany (A) | Mean | 1.86 | 16.86 | 17.88 |
| | N | 91 | 51 | 50 |
| | Minimum | 0 | 0 | 0 |
| | Maximum | 8 | 70 | 70 |
| | SD | 2.1 | 15.52 | 15.76 |
| | Greater than | - | - | - |
| Hungary (B) | Mean | 1.75 | 37.82 | 25.73 |
| | N | 84 | 54 | 52 |
| | Minimum | 0 | 0 | 3 |
| | Maximum | 9 | 100 | 65 |
| | SD | 2.03 | 27.67 | 16.62 |
| | Greater than | - | AD | AC |
| Netherlands (C) | Mean | 2.22 | - | 15.41 |
| | N | 93 | - | 47 |
| | Minimum | 0 | - | 0 |
| | Maximum | 10 | - | 50 |
| | SD | 2.38 | - | 13.29 |
| | Greater than | - | - | - |
| Poland (D) | Mean | 2 | 18.41 | 20.25 |
| | N | 156 | 99 | 101 |
| | Minimum | 0 | 0 | 0 |
| | Maximum | 10 | 100 | 70 |
| | SD | 2.22 | 20.56 | 14.34 |
| | Greater than | - | - | - |
| Spain (E) | Mean | 1.72 | 49.61 | 44.52 |
| | N | 46 | 33 | 33 |
| | Minimum | 0 | 33 | 33 |
| | Maximum | 8 | 100 | 100 |
| | SD | 1.67 | 20.48 | 16.38 |
| | Greater than | - | ABD | ABCD |

**Table C.5.** *Correlation between share of land affected and share of total farm revenue affected.*

| | | | % of cultivated land affected by the most critical yield loss | % of total farm revenue affected by the most critical yield loss |
|---|---|---|---|---|
| Kendall's tau_b | | | | |
| | % of cultivated land affected by the most critical yield loss | Correlation coefficient | 1.000 | 0.397(**) |
| | | Sig. (1-tailed) | | 0.000 |
| | | N | 693 | 680 |
| | % of total farm revenue affected by the most critical yield loss | Correlation coefficient | 0.397(**) | 1.000 |
| | | Sig. (1-tailed) | 0.000 | |
| | | N | 680 | 684 |
| Spearman's rho | | | | |
| | % of cultivated land affected by the most critical yield loss | Correlation coefficient | 1.000 | 0.508(**) |
| | | Sig. (1-tailed) | | 0.000 |
| | | N | 693 | 680 |
| | % of total farm revenue affected by the most critical yield loss | Correlation coefficient | 0.508(**) | 1.000 |
| | | Sig. (1-tailed) | 0.000 | |
| | | N | 680 | 684 |

** Correlation is significant at the 0.01 level (1-tailed).

**Table C.6.** *Correlation between share of livestock production affected and share of total farm revenue affected.*

| | | | % of livestock affected by the most critical production loss | % of total farm revenue affected by the most critical production loss |
|---|---|---|---|---|
| Kendall's tau_b | | | | |
| | % of livestock affected by the most critical production loss | Correlation coefficient | 1.000 | 0.502(**) |
| | | Sig. (1-tailed) | | 0.000 |
| | | N | 237 | 231 |
| | % of total farm revenue affected by the most critical production loss | Correlation coefficient | 0.502(**) | 1.000 |
| | | Sig. (1-tailed) | 0.000 | |
| | | N | 231 | 236 |
| Spearman's rho | | | | |
| | % of livestock affected by the most critical production loss | Correlation coefficient | 1.000 | 0.624(**) |
| | | Sig. (1-tailed) | | 0.000 |
| | | N | 237 | 231 |
| | % of total farm revenue affected by the most critical production loss | Correlation coefficient | 0.624(**) | 1.000 |
| | | Sig. (1-tailed) | 0.000 | |
| | | N | 231 | 236 |

** Correlation is significant at the 0.01 level (1-tailed).

**Table C.7.** *Farmers' ratings of methods useful for reducing risk of yield loss (member state averages; 1-3- Not effective, 3-5-Moderately effective, 5-7-Very effective).*

| | | Crop rotation/ relay planting | Irrigation | Drainage | Preventive plant protection | Technological improvement |
|---|---|---|---|---|---|---|
| Germany (A) | Mean | 5.64 | 3.62 | 3.66 | 5.93 | 5.08 |
| | N | 191 | 183 | 175 | 189 | 192 |
| | SD | 1.03 | 1.95 | 1.68 | 1.1 | 1.2 |
| | Greater than | BDE | - | D | E | - |
| Hungary (B) | Mean | 4.65 | 4.26 | 4.06 | 5.53 | 4.78 |
| | N | 191 | 173 | 179 | 192 | 187 |
| | SD | 1.74 | 2.03 | 1.96 | 1.53 | 1.32 |
| | Greater than | E | A | D | E | - |
| Netherlands (C) | Mean | 5.1 | 4.48 | 5.12 | 5.8 | 4.97 |
| | N | 82 | 75 | 76 | 84 | 72 |
| | SD | 2.06 | 2.08 | 2.05 | 1.5 | 1.7 |
| | Greater than | E | A | ABD | E | - |
| Poland (D) | Mean | 4.62 | 5.53 | 3.04 | 5.45 | 4.92 |
| | N | 202 | 200 | 201 | 202 | 202 |
| | SD | 1.8 | 1.8 | 1.87 | 1.54 | 1.7 |
| | Greater than | E | ABC | - | E | - |
| Spain (E) | Mean | 3.7 | 5.64 | 4.74 | 4.77 | 5.01 |
| | N | 142 | 144 | 140 | 141 | 143 |
| | SD | 2.05 | 1.42 | 1.8 | 1.77 | 1.64 |
| | Greater than | - | ABC | ABD | - | - |

**Table C.8.** *Farmers' ratings of methods useful for reducing risk of livestock loss (member state averages; 1-3- Not effective, 3-5-Moderately effective, 5-7-Very effective).*

| | | Preventive medical treatment | Ex-post medical treatment | Quarantines and building rotation | Young animals from own breeding | Quality assurance |
|---|---|---|---|---|---|---|
| Germany (A) | Mean | 4.47 | 4.49 | 4.13 | 5.19 | 5.43 |
| | N | 103 | 100 | 99 | 97 | 98 |
| | SD | 1.49 | 1.2 | 1.71 | 1.89 | 1.64 |
| | Greater than | - | - | C | D | BCD |
| Hungary (B) | Mean | 5.52 | 4.48 | 3.76 | 4.9 | 3.89 |
| | N | 87 | 86 | 84 | 83 | 83 |
| | SD | 1.62 | 1.4 | 1.64 | 1.73 | 1.68 |
| | Greater than | AC | - | C | - | - |
| Netherlands (C) | Mean | 4.03 | 5.48 | 2.79 | 5.09 | 4.34 |
| | N | 118 | 122 | 100 | 120 | 110 |
| | SD | 2 | 1.45 | 1.92 | 2.1 | 1.98 |
| | Greater than | - | AB | - | D | - |
| Poland (D) | Mean | 5.95 | 5.34 | 3.95 | 4.51 | 4.74 |
| | N | 164 | 164 | 164 | 163 | 164 |
| | SD | 1.42 | 1.59 | 1.74 | 1.92 | 1.81 |
| | Greater than | AC | AB | C | - | B |
| Spain (E) | Mean | 5.82 | 5.55 | 5.41 | 5.04 | 5.47 |
| | N | 49 | 49 | 49 | 48 | 49 |
| | SD | 1.19 | 1.28 | 1.49 | 1.66 | 1.49 |
| | Greater than | AC | AB | ABCD | - | BC |

**Table C.9.** *Marketing channels applied by farmers for selling farm products.*

| | How do farmers sell the majority of their products? | | | | | | | | |
|---|---|---|---|---|---|---|---|---|---|
| | Individual sales | | Through cooperative | | Through marketing contract | | Through production contract | | Total |
| Germany; n=199 (A) | 41.7% | C | 40.7% | BD | 13.6% | E | 4.0% | - | 100% |
| Hungary; n= 200 (B) | 70.0% | ACE | 9.5% | - | 16.5% | CE | 4.0% | - | 100% |
| Netherlands; n=222 (C) | 21.2% | - | 64.9% | ABD | 7.2% | - | 6.8% | E | 100% |
| Poland; n=206 (D) | 60.2% | ACE | 13.1% | - | 19.9% | CE | 6.8% | E | 100% |
| Spain; n=200 (E) | 43.5% | C | 53.0% | BD | 3.0% | - | 0.5% | - | 100% |

**Table C.10.** *Availability of off-farm revenue among farmers.*

| | Does the respondent have off-farm revenue? | | | | |
|---|---|---|---|---|---|
| | Yes | | No | | Total |
| Germany; n=201 (A) | 60.7% | BCD | 39.3% | - | 100% |
| Hungary; n= 204 (B) | 38.7% | - | 61.3% | AE | 100% |
| Netherlands; n=236 (C) | 41.1% | - | 58.9% | A | 100% |
| Poland; n=206 (D) | 30.1% | - | 69.9% | AE | 100% |
| Spain; n=200 (E) | 53.5% | BD | 46.5% | - | 100% |

**Table C.11.** *Average proportion of off-farm revenue within farmers' total revenue across countries (where there is off-farm revenue).*

| | | % of total revenues from off-farm activities in 2005 |
|---|---|---|
| Germany (A) | Mean | 34.19 |
| | N | 122 |
| | Minimum | 1 |
| | Maximum | 95 |
| | SD | 25.18 |
| | Greater than | - |
| Hungary (B) | Mean | 38.26 |
| | N | 79 |
| | Minimum | 1 |
| | Maximum | 95 |
| | SD | 23.86 |
| | Greater than | - |
| Netherlands (C) | Mean | 31.89 |
| | N | 97 |
| | Minimum | 1 |
| | Maximum | 100 |
| | SD | 29.27 |
| | Greater than | - |
| Poland (D) | Mean | 33.32 |
| | N | 62 |
| | Minimum | 1 |
| | Maximum | 90 |
| | SD | 21.57 |
| | Greater than | - |
| Spain (E) | Mean | 61.22 |
| | N | 107 |
| | Minimum | 2 |
| | Maximum | 100 |
| | SD | 24.4 |
| | Greater than | ABCD |

**Table C.12.** *Distribution of off-farm revenue during the year (if such exists).*

| | Is the off-farm revenue year round or just seasonal? | | | | |
|---|---|---|---|---|---|
| | Year round | | Seasonal | | Total |
| Germany; n=129 (A) | 77.5% | - | 22.5% | - | 100% |
| Hungary; n= 104 (B) | 75.0% | - | 25.0% | E | 100% |
| Netherlands; n=119 (C) | 81.5% | - | 77.3% | - | 100% |
| Poland; n=75 (D) | 18.5% | - | 22.7% | - | 100% |
| Spain; n=108 (E) | 89.8% | B | 10.2% | - | 100% |

**Table C.13.** *Existence of debt towards bank(s) at the time of completing the questionnaire.*

| | Do farmers have debt at the moment? | | | | |
|---|---|---|---|---|---|
| | Yes | | No | | Total |
| Germany; n=201 (A) | 29.4% | - | 70.6% | CD | 100% |
| Hungary; n= 204 (B) | 22.1% | - | 77.9% | CD | 100% |
| Netherlands; n=236 (C) | 54.2% | ABE | 45.8% | - | 100% |
| Poland; n=206 (D) | 65.0% | ABE | 35.0% | - | 100% |
| Spain; n=200 (E) | 18.0% | - | 82.0% | CD | 100% |

**Table C.14.** *Farmers' perceptions of credit access adequacy.*

| | How do farmers perceive the adequacy of access to credit capacity? | | | |
|---|---|---|---|---|
| | There is no access to credit at all | | There is timely access but only with hard conditions and high costs | |
| Germany; n=191 (A) | 2.6% | - | 11.0% | - |
| Hungary; n=187 (B) | 27.3% | ACDE | 54.5% | ACD |
| Netherlands; n=214 (C) | 3.7% | - | 9.8% | - |
| Poland; n=206 (D) | 1.0% | - | 24.3% | AC |
| Spain; n=179 (E) | 8.9% | D | 50.3% | ACD |

**Table C.15.** *Application of quality assurance systems.*

| | Do farmers apply any quality assurance systems? | | | | |
|---|---|---|---|---|---|
| | Yes | | No | | Total |
| Germany; n=200 (A) | 90.0% | BDE | 10.0% | - | 100% |
| Hungary; n=198 (B) | 20.2% | - | 79.8% | ACDE | 100% |
| Netherlands; n=218 (C) | 81.7% | BD | 18.3% | - | 100% |
| Poland; n=206 (D) | 68.4% | B | 31.6% | AC | 100% |
| Spain; n=200 (E) | 74.5% | B | 25.5% | A | 100% |

**Table C.16.** *Recent participation in any farming related professional educational program.*

| | Do farmers participate in any professional educational program related to farming recently? | | | | |
|---|---|---|---|---|---|
| | Yes | | No | | Total |
| Germany; n=196 (A) | 76.0% | BCDE | 24.0% | - | 100% |
| Hungary; n=194 (B) | 15.5% | - | 84.5% | ACD | 100% |
| Netherlands; n=228 (C) | 61.8% | BDE | 38.2% | A | 100% |
| Poland; n=206 (D) | 30.6% | B | 69.4% | AC | 100% |
| Spain; n=193 (E) | 18.7% | - | 81.3% | AC | 100% |

| Costs and conditions are reasonable but requires a long procedure | | There is timely access with reasonable conditions and costs | | Total |
|---|---|---|---|---|
| 8.4% | - | 78.0% | BDE | 100% |
| 11.8% | - | 6.4% | - | 100% |
| 5.1% | - | 81.3% | BDE | 100% |
| 41.3% | ABCE | 33.5% | B | 100% |
| 14.0% | C | 26.8% | B | 100% |

**Table C.17.** *Application of additional health insurance besides the obligatory one.*

| | Do farmers have additional health insurance (besides the obligatory one)? | | | | |
|---|---|---|---|---|---|
| | Yes | | No | | Total |
| Germany; n=192 (A) | 52.6% | BDE | 47.4% | C | 100% |
| Hungary; n=202 (B) | 16.8% | - | 83.2% | AC | 100% |
| Netherlands; n=224 (C) | 70.5% | ABDE | 29.5% | - | 100% |
| Poland; n=206 (D) | 17.0% | - | 83.0% | AC | 100% |
| Spain; n=174 (E) | 22.4% | - | 77.6% | AC | 100% |

**Table C.18.** *Application of life insurance products.*

| | Do farmers have life insurance? | | | | |
|---|---|---|---|---|---|
| | Yes | | No | | Total |
| Germany; n=195 (A) | 92.8% | BCDE | 7.2% | - | 100% |
| Hungary; n=202 (B) | 29.7% | - | 70.3% | AC | 100% |
| Netherlands; n=225 (C) | 62.2% | BDE | 37.8% | A | 100% |
| Poland; n=206 (D) | 29.6% | - | 70.4% | AC | 100% |
| Spain; n=175 (E) | 42.3% | - | 57.7% | AC | 100% |

# Appendix D

## Whole-farm modelling assumptions

**Table D.1.** *Number of farms per ESU[1] cluster and typology in five member states[2].*

| | ESU | Farm typology (Eurostat)[3] | |
|---|---|---|---|
| | | *13* Specialist cereals, oilseed and protein crops | *14* General field cropping |
| Germany | *Total, >0* | *58,160* | *31,490* |
| | ≥8 and <16 | 12,060 | 4,310 |
| | ≥16 and <40 | 9,040 | 6,690 |
| | ≥40 and <100 | 5,390 | **7,740** |
| | ≥100 | 4,670 | 6,280 |
| Hungary | *Total, >0* | *74,840* | *33,420* |
| | ≥8 and <16 | **4,530** | 1,440 |
| | ≥16 and <40 | 3,540 | 980 |
| | ≥40 and <100 | 1,490 | 290 |
| | ≥100 | 620 | 250 |
| Netherlands | *Total, >0* | *1,250* | *11,050* |
| | ≥8 and <16 | 380 | 1,660 |
| | ≥16 and <40 | 190 | 2,540 |
| | ≥40 and <100 | 110 | **3,160** |
| | ≥100 | 50 | 2,310 |
| Poland | *Total, >0* | *283,370* | *410,540* |
| | ≥8 and <16 | **4,180** | 9,940 |
| | ≥16 and <40 | 2,860 | 4,400 |
| | ≥40 and <100 | 1,310 | 960 |
| | ≥100 | 630 | 480 |
| Spain | *Total, >0* | *123,790* | *45,080* |
| | ≥8 and <16 | 21,110 | 8,290 |
| | ≥16 and <40 | **22,420** | 10,390 |
| | ≥40 and <100 | 9,830 | 5,100 |
| | ≥100 | 2,040 | 2,600 |

[1]European Size Unit.

[2]Per member state farms representing bold cells are selected for in-depth risk analysis.

[3]Typology code and description.

**Table D.2.** *Variable cost per member state.*

| | Variable costs (euro/ha)[1] | | | | |
|---|---|---|---|---|---|
| | Germany | Hungary | Netherlands | Poland | Spain |
| Wheat | 700 | 162 | 500 | 300 | 400 |
| Rye | 580 | 66 | 400 | 190 | 300 |
| Barley | 630 | 126 | 350 | 230 | 350 |
| Oats | 480 | 77 | 300 | 200 | 300 |
| Summer cereals | 598 | 108 | 388 | 230 | 338 |
| Maize | 1,150 | 275 | 525 | 390 | 1,200 |
| Cereals | 598 | 108 | 388 | 230 | 338 |
| Potatoes | 2,780 | 799 | 2,000 | 750 | 200 |
| Sugar beet | 1,148 | 486 | 850 | 620 | 1,500 |
| Rapeseed | 710 | 194 | 600 | 380 | NF |
| Sunflower | NF[2] | 184 | NF | NF | 400 |

[1]Seed, fertilizers, pesticides, energy, irrigation, other crop production costs.

[2]Not a feasible crop considered in the respective member state.

**Table D.3.** *Constraints in the model per member state.*

| Crops | Germany | | Hungary | | Netherlands | | Poland | | Spain | |
|---|---|---|---|---|---|---|---|---|---|---|
| | Obs[1] | M[2] | Obs | M | Obs | M | Obs | M | Obs | M |
| Wheat | 32% | 0.67 | 30% | 0.67 | 19% | 1.00 | 40% | 0.50 | 12% | 0.30 |
| Rye | 3% | 0.67 | 0% | 0.67 | 0% | 1.00 | 10% | 0.67 | 1% | 0.05 |
| Barley | 16% | 0.67 | 6% | 0.67 | 6% | 1.00 | 13% | 0.50 | 44% | 0.60 |
| Oats | 1% | 0.33 | 1% | 0.67 | 1% | 1.00 | 8% | 0.67 | 2% | 0.05 |
| Summer cereals | 0% | 0.00 | 0% | 0.67 | 0% | 0.00 | 4% | 0.00 | 0% | 0.00 |
| Maize | 2% | 0.33 | 29% | 0.67 | 0% | 1.00 | 2% | 0.10 | 5% | 0.10 |
| Cereals | 2% | 0.00 | 4% | 0.67 | 0% | 0.00 | 6% | 0.00 | 0% | 0.00 |
| Potatoes | 5% | 0.20 | 0% | 0.00 | 21% | 0.33 | 1% | 0.02 | 0% | 0.01 |
| Sugar beet | 16% | 0.25 | 0% | 0.00 | 14% | MO[4] | 2% | MO | 1% | 0.01 |
| Rapeseed | 4% | 0.25 | 2% | 0.20 | 0% | 1.00 | 1% | 0.02 | 0% | 0.00 |
| Sunflower | 0% | 0.00 | 16% | 0.00 | 0% | 0.00 | 13% | 0.00 | 7% | 0.00 |
| Other crops | 20% | Obs[3] | 13% | Obs | 39% | Obs | 1% | Obs | 28% | Obs |

[1]Observed production plan.
[2]Maximum allowed in production plan as fraction of total farm land.
[3]Treated as an equality instead of an inequality.
[4]Maximum is set at observed level.

# Index